Comets, Meteors and Asteroids

Comets, Meteors and Asteroids

John Man

First published in 2001 by
BBC Worldwide Ltd,
Woodlands, 80 Wood Lane,
London W12 0TT

© BBC Worldwide Ltd 2001

DK PUBLISHING, INC.
www.dk.com

Publisher: Sean Moore
Art Director: Dirk Kaufman
Editorial Director: Chuck Wills

First American Edition, 2001

00 01 02 03 04 05 10 9 8 7 6 5 4 3 2 1

Published in the United States by
DK Publishing, Inc.
95 Madison Avenue
New York, New York 10016

ISBN 0-7894-8159-6

Produced for BBC Worldwide by
Toucan Books Ltd, London

Cover photograph: John
Thomas/Science Photo Library

Printed and bound in France by
Imprimerie Pollina s.a.
Color separation by Imprimerie
Pollina s.a.

Contents

THE REMNANTS OF CREATION

THE REMNANTS OF CREATION

In ancient times, people lived close to their gods, and the gods made the heavens resound. According to the Latin writer Lucretius, when Zeus fought the Titans, 'thick and fast, the thunderbolts, with thunder and lightning, flew from his stout hand'. In Norse mythology, the twilight of the gods was marked by fire on Earth. Once such tales seemed mere superstition, but not any more. It seems that the ancients knew something we had forgotten until recently: that the heavens can rain fire – meteorites in today's terminology. But rocks that strike the Earth are a minute percentage of the small bodies that wander between the planets. And they are much more than destroyers. Born in the remotest times and regions, these wanderers – comets, meteors and asteroids, as well as meteorites – reveal how our Solar System and our world came into being. They have been part of our past, and will, sooner or later, play a role in our future.

Previous page: The early Solar System's disc of dust collapses into rings. In each ring, dust gathers into asteroids and planets – like the one lower left – and these are constantly at risk from comets.

ORIGINS

At one time, it was taken for granted that the heavens determined human destinies. Scientific advances over the last two centuries rejected such notions as mere superstition, assuring us that our Solar System was a safe and stable place in which planets, moons and comets circled according to Newton's neat laws, remote from human life. Now scientists have come to realize that we are, after all, bound to the heavens – in particular to the little objects that swing randomly between the planets.

To understand the nature of these objects, go back in your mind's eye to a time before the Earth existed. Some 4600 million years ago, towards the edge of our galaxy of 100,000 stars, a tenuous cloud of interstellar gas and dust was blasted out by the cataclysmic explosion of a nearby star. This particular region of the young galaxy, with its scattering of hydrogen and helium molecules, had been maturing for some 10,000 million years, steadily enriched by elements created and ejected in previous, more distant stellar explosions. At some point, a random meeting of dust and gas produced a tiny particle that was minutely denser than its surroundings. That fractional difference was enough for gravity to act. Steadily, it drew in nearby wisps of gas and dust, became roughly circular and began to collapse in upon itself.

The temperature began to rise from near absolute zero (-273°C) to 1000°C. Plumes of gas carried the excess heat to what was now becoming a surface, where the gas cooled, giving off a dull glow, before gravity dragged it back towards the

1. After 1000 million years, the young Solar System was an unstable spiral of proto-planets, planetesimals, asteroids and comets.

depths. The gas-ball's initial slight movements became a spin, which increased in speed as the ball shrank, in much the same way as a spinning skater's speed increases as she draws in her arms. After about 50 million years, the core of the disc hit a temperature of 8 million °C. The hydrogen ignited and the Sun was born.

Meanwhile, the shrinking centre of the gas-ball left behind swirls of gas and dust, rather like the outer regions of a whirlpool. Under the combined effects of gravity and centrifugal force, these swirls flattened into a disc, with the inner and outer areas moving at different speeds, breaking up and then reforming into lesser swirls. Since any grain of dust even a few metres farther from the Sun orbited at a slightly slower speed, the inner grains overtook the outer ones. Drawn by gravity, they collided, sometimes forming small flakes, then ever larger objects, from bodies the size of gravel, to pebbles, rocks, and finally mountain-sized chunks that are known

as 'planetesimals'. Sometimes the collisions were so great that they smashed the forming rocks apart. Computer simulations suggest that finally – after perhaps 100 million years of accretion, collision, fracture and reformation – the accumulated material was assembled into the rough cores of the nine planets we know today.

Formation of the planets

The formation of the planets varied with distance from the young Sun's blast of radiant energy. Close in, where the temperature was already perhaps 2000°C, it was too hot for the grains to stick together. Between about 80–320 million km (50–200 million miles) out, where the temperature fell to about 300°C, gases would still be kept in unrestrained motion by the heat, but solid grains of dust could stick together in ever-growing lumps. Four of these lumps survived, forming the cores of

 THE LARGEST IMPACT CRATER

One of Jupiter's moons, Callisto (right), is marked by the Solar System's largest impact crater. Known as Valhalla, the crater's central plain covers an area about the size of Germany. It is bright with ice, and looks like a bull's eye, with 30 target-like rings of distorted rock rolling out for some 2600 km (1600 miles), dominating an entire hemisphere, and large enough to span the whole of Europe. The crater was formed by a stupendous collision that sent melted rock and ice flooding outwards, only to be quick-frozen in the surface temperature of -165°C.

the inner planets: Mercury, Mars, Earth and Venus. Most of the remaining gases were blown away from the inner planets by the solar wind, the stream of radiation emitted by the Sun. Under this assault, and now under the influence of their own trapped radioactive heat, the cores melted and condensed. Less dense materials rose to the surface to form a mantle and an outer skin, or crust.

In the colder outer regions of the disc, too far out for the Sun to raise the temperature of the gases much above absolute zero, elemental grains and gas mixed to form the cores of what would become four giant gaseous planets: Jupiter, Saturn, Neptune and Uranus. These steadily swept the interplanetary spaces clear of leftover detritus. (Pluto, the most distant planet, remains an oddity – its eccentric orbit and tiny size suggest that it may possibly be an escaped moon.)

But not all the detritus was used up: the remainder – solid lumps in the inner Solar System, icy balls in the outer regions – forms the subject of this book.

1. The rings of Saturn are made of dust and pebbles sorted into stable bands by collisions and the effects of gravity. They suggest a small-scale model of how the Solar System formed.

1

THE GREAT BOMBARDMENT

The formation theory was, in essence, a process suggested by the philosopher Immanuel Kant in the mid-18th century. He was speculating, but today's astronomers know that it is accurate in general terms because they can see the dust-veils and forming planetary discs around other stars. Although details remain mysterious, the theory explains many things about our Solar System. For instance, it explains why the planets all orbit in the same direction, which is also the same direction as the Sun's spin, and why they are all on the same plane. (If the Solar System were the size of a pancake, it would be only 1 cm [1/2 in] thick, with all the planets' orbits contained within it.) It also explains why all the major planets have almost circular orbits, and why the inner planets are small and solid, while the major outer planets (except for Pluto) are gaseous giants.

Asteroid, meteorite or comet?

The theory also explains the existence of a vast mass of unused material – dust particles, rocks and drifting gas. The constituents of this material are known by a variety of names, depending partly on their nature, and partly on how they are observed by us on Earth. Dust particles – typically, grains that are less than a micron (one-thousandth of a millimetre) across – are too small to be seen by telescope, or even in the palm of your hand, but

1

1. Interstellar dust – the stuff of comets – consists of grains that combine into particles like this one, 10 microns (one-hundredth of a millimetre) across.

2. Meteors often break up, giving the false impression of random bombardment. This 1719 woodcut portrays a fearful mixture of stars, comets and meteors.

3. This record of a meteor breaking up in 1783 reveals a more accurate view.

2

they are visible enough when bunched into interstellar clouds and, on any clear night, when they zip into the atmosphere and burn up as meteors. (The name 'meteor' derives from the Greek word covering all atmospheric phenomena, the study of which was termed 'meteorology'. It was not until the 19th century that meteorology became restricted to the study of weather.)

The heavier bodies, some in regular orbits, some swinging wildly between the planets, are known as asteroids. An asteroid can be any size, from a stone to a planetesimal. If an asteroid survives the searing journey through the atmosphere and strikes the Earth, it is known as a meteorite. Near the Sun, lightweight gaseous balls are blasted out of existence, or driven by the pressure of radiation far into the outer reaches of the Solar System and beyond. There, the gas mixes with dust to form the raw material from which comets are made. Only when these objects approach the Sun do they take

3

> ☆ In Europe the year AD 902 was nicknamed 'Year of the Stars' because meteors streaked across the sky as thickly as snowflakes.

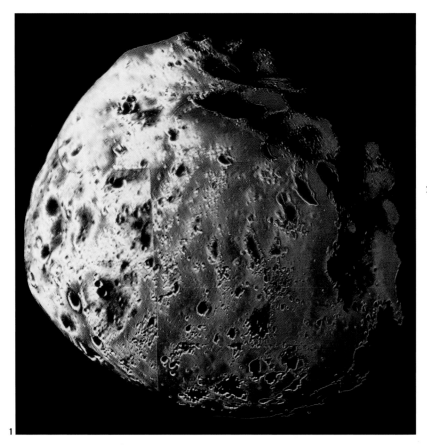

1. Circling Jupiter, Ganymede is the Solar System's largest satellite. Varied terrains with scattered small craters suggest that geological activity destroyed larger, older craters.

2. Phobos, one of Mars's two tiny moons, is probably a captured asteroid. Its many craters include one made by an impact that almost shattered it.

the shape that has been familiar to humans from early times, with their glowing heads and the streaming tails from which they derive their Greek name *aster kometes* (long-haired star).

Once, these categories seemed to define quite separate objects, different also from moons and planets. Were not asteroids made of inert rock and iron, comets of soft material made active by

sunlight? Now astronomers know that all these categories can merge into each other. A comet stripped of its gas may become an asteroid – currently, astronomers estimate that about one-third of asteroids are 'dead' comets. Both are made up of gas and dust, to which they return if they split up. A moon may be a captured asteroid (as the two little moons of Mars probably are), and an asteroid an escaped moon. A moon may be larger than a planet (as Jupiter's Ganymede is larger than Mercury). A large asteroid may be termed a minor planet, and vice versa: Pluto, commonly called a planet, is in such an eccentric orbit that it might be better described as an asteroid; on the other hand, it has its own little moon, Charon, which just happens to be the size of the largest known asteroid, Ceres.

Evolution of the Solar System

The detritus of rock, dust and gas is of particular significance for the evolution of the Solar System, as the process of formation and reformation has never ceased. The Sun and the planets continued to sweep up the leftover bits. Some of the rubbish fell into the Sun. For the planets, a slow-motion record of their first few hundred million years would have

 THE MARS ROCK

In 1984 meteorite hunters searching the Allan Hills, Antarctica, found a greenish rock the size of a potato. Named ALH 84001 — the first rock of the year from Allan Hills — the meteorite had an astonishing history, as revealed by its chemical structure and little pockets of air sealed in 4500 million years ago. A comparison with the atmosphere collected by the Viking lander in 1976 showed that the rock was formed on Mars, where it was tossed by the force of an asteroid or cometary impact into water. It lay for millions of years as Mars dried out, was blasted into space by another impact 15 million years ago, and finally landed in the Antarctic some 13,000 years ago. A team of NASA researchers headed by David S. McKay found microscopic shapes that looked like micro-fossils, suggestive of primitive Martian microbes. The 1996 announcement of possible evidence of life on Mars sparked international interest, and a scientific furore. No scientists claim that the evidence is foolproof, but few reject the possibility outright. Whatever the conclusion, the Mars Rock intensified research into meteorites, Mars and the nature of life.

The Mars Rock found in Antarctica in 1984.

Mars micro-fossils enlarged 100,000 times.

been filled with drama and catastrophe, with each planet subject to an irregular bombardment, each new impact blasting a new crater, sometimes splashing molten rock far into space, even side-swiping smaller planets so that their axes were tilted or their spin reversed (which is why three planets – Venus, Uranus and Pluto – rotate in the opposite direction to their movement round the Sun).

Only after about 600 million years did the Solar System settle into something like stability. Those bodies with an active atmosphere and geology reformed their own surfaces, but many still show evidence of the bombardment that ended some 4000 million years ago. The Moon is almost certainly the product of that early bombardment, when an asteroid struck the Earth a glancing blow, blasting much of the young Earth's mantle into debris that formed our satellite. As the Moon's surface hardened, more debris fell, melting rock

and creating the huge lunar plains still known by the Latin name *maria* (seas). Gradually, the rain of rock became more scattered, and the objects smaller, falling from more distant parts.

Mercury suffered a similar fate, and looks very like the Moon. Caloris Basin on Mercury, a huge, lava-filled feature some 1300 km (800 miles) across, resembles a lunar *mare* (sea). Like the Moon, Mercury has retained a regular scattering of later, smaller craters, but fewer of them – a difference that suggests the distribution of meteorites was modified by the Sun's gravitational field. On Mars, craters are scattered unevenly. Apparently, the northern hemisphere of Mars was more subject to volcanoes and lava outflows that covered it in craters. The southern hemisphere resembles the Moon and Mercury, though many of the craters are filled with wind-blown sand dunes. On Venus, a dense atmosphere, vicious winds and volcanoes reformed

1. The Moon's dark 'seas' are the filled-in remnants of immense impacts. These were possibly caused by material raining down after the Moon was blasted apart from the Earth.

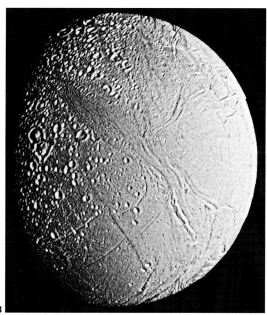

3

2. On Mercury, smaller, later craters scar Caloris, a 'sea' of molten rock made by an ancient impact.

3. Saturn's moon Enceladus has craters, but also plains made by later flows of ice and lava.

2

the surface, but several massive craters survived. Ganymede, Jupiter's largest moon, and its sister, Callisto, are both pockmarked with craters. All but one (Titan) of Saturn's 18 known moons are cratered. Of the 15 known moons of Uranus, several have cratered features; in fact, on Umbriel the craters are so densely packed that they overlap.

Over the course of 4000 million years, most of the dust and the larger bodies have been blotted up, but there is still a good deal of both left, falling as a sort of 'hard rain' on all planets. Most of the rain consists of droplets of dust particles, which arrive in huge numbers, perhaps as many as 100 million each day on Earth. Most are smaller than pinheads, and burn up unseen. Only the larger ones penetrate farther, and vaporize as meteors. Anyone watching on a clear night would expect to see about half a dozen an hour. Regularly, the circling specks arrive in swarms, creating 'meteor showers', of which there are a dozen or so major ones each year.

PROFILE OF A COMET

For the most part, hard rain comes from comets, which are in a continuous state of disintegration. The head consists of loosely bound icy grains, forming what astronomers often refer to as a 'dirty snowball', a term coined by the American astronomer Fred Whipple in 1950. For much of the time, most of these snowballs circle in the icy interstellar wastes, so far from the Sun that they would appear to an astronaut as not much brighter than any other star. Only when the snowballs fall towards the Sun do they become true comets. Drawn by the Sun's gravity, their gases heat up and erupt as they accelerate inwards. The tail, boiled off by the Sun's heat, is so ethereal that the pressure of the Sun's radiation – the solar wind – ensures that it always points away from the Sun, even as the comet swings round the Sun in a tight turn to vanish outwards again.

2

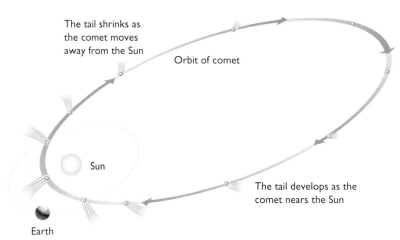

The tail shrinks as the comet moves away from the Sun

Orbit of comet

Sun

Earth

The tail develops as the comet nears the Sun

1. Most comets have elongated orbits, often stretching far beyond the orbit of Pluto. Tails are formed by the Sun's heat and blown outwards by the solar wind of light and radiation.

1

2. A meteor burns up in the night sky. Visible for less than a second, it was caught on camera through a lens left open so that the stars appear blurred.

3. Behind a foreground of stars blurred by a long exposure, the 1992 comet Swift-Tuttle displays the fuzzy head and tail that are a comet's main traits.

Short-period comets circle close in to the Sun, returning in no more than 200 years. Some interact with a planet or two, and are slung for ever out of the Solar System. Long-period comets may take up to 10 million years to make a single circuit of the Sun, ranging out halfway to the nearest star, in a 'sink' where dormant comets are assumed to lurk by the billion.

Comets are a double source of hard rain. Their tenuous tails form extended clouds of interstellar dust, and remain in position for decades, drifting and diffusing slowly. Many tails intersect the Earth's orbit. When the Earth passes through them, they are the stuff of meteor showers: the Perseid shower

is linked to the comet known as Swift-Tuttle, discovered in the 19th century and rediscovered in 1992. Swift-Tuttle comes by once every 130 years, but its debris remains as a sort of fossilized wake so that the Earth sweeps through it every year.

Comets can be torn apart by the influence of a planet's gravity and swept to dramatic ends. In 1826, for instance, a new comet was found with a short period of 6.75 years. By 1845 – three passes later – it had broken in two. In 1872 some of it fell to Earth in a spectacular one-off meteor shower. The most impressive break-up occurred in the early 1990s, when a new comet was found and named Shoemaker-Levy, after the two astronomers who identified it. The comet, caught by Jupiter's gravitational field, broke up into an extended line of sub-comets that finally fell into Jupiter in September 1994 with a series of massive explosions (▷ p. 86). Only if the core survives after its volatile gases are blown away by multiple passes round the Sun does the comet evolve into an asteroid, like millions of others in orbit between the planets.

Asteroids in close-up

Asteroids, which range in size from irregular large rocks to roughly spherical objects the size of small moons, orbit the Sun in uncounted thousands. Recently, they have come in for ever more detailed analysis from ground-based telescopes, radar and spacecraft.

Twenty-five of the known asteroids are more than 100 km (62 miles) across, and another 50 or so are 75–100 km (47–62 miles) across. Below this size, the asteroids increase in number as they decrease in diameter. Some 8000 are named, and all are numbered. (Astronomers usually add numbers to the names: 253 Mathilde, 288 Glauke.) More than 30,000 asteroids, typically about 1 km (0.6 miles) across, are now known but this is only a tiny percentage of the millions that must exist, down to rocks, pebbles and gravel-sized grains.

Most asteroids occupy the Asteroid Belt between Mars and Jupiter (▷ Chapter 2), the raw material, perhaps, of a planet that was prevented from forming by the disruptive influence of Jupiter's massive gravitational field. In this privileged position, they are protected from further

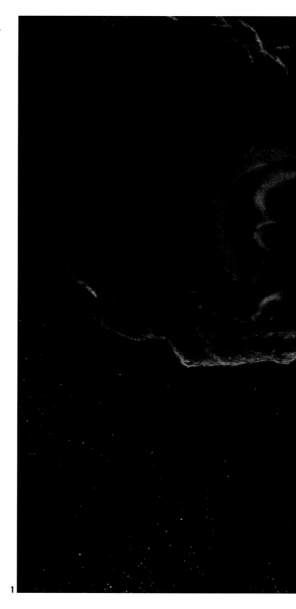

1

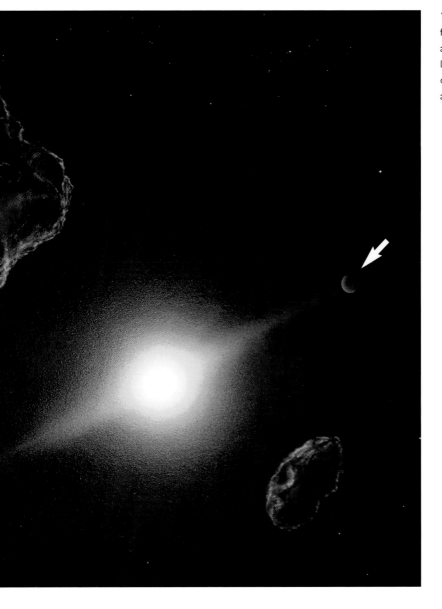

1. In this imagined view from the Asteroid Belt, two asteroids circle the Sun, its light scattered by a local dust-cloud. The planet arrowed is Mars.

gravitational interaction with a planet, but not from each other. Although millions of asteroids easily fit into their allotted space – several probes have passed through the Asteroid Belt without mishap – their orbits vary, rather like dodgem cars moving in three dimensions. In the long run, every one of them will be involved in numerous collisions.

For aeons, perhaps, an ancient asteroid is merely cratered by dust and pebbles, but then eventually something large strikes the asteroid, at a speed of anything up to 5 km (3 miles) per second, enough to shatter any object smaller than several kilometres across. Often, smaller boulders produced by earlier collisions rejoin, creating irregular bits of rubble. One, Castalia, is a 'double' asteroid made up of two boulders spinning once every four hours, while another, Geographos, is a cigar-shaped asteroid 5 km (3 miles) long but less than 2 km (1$1/4$ miles) across. The result of this chaotic history is that asteroids are very varied objects, some preserving the chemistry of the early Solar System, others having been battered and melted, turning them into chemical cauldrons like miniature planets.

The Asteroid Belt is not the only home for these inert lumps. Others, known as Trojans, circle in Jupiter's orbit, in two groups moving in a sort of gravitational harmony 60 degrees in front of and behind the giant planet. A few asteroids swing in eccentric orbits, some crossing the Earth's orbit and occasionally approaching dangerously near. Like the planets, they almost all fall within the plane of the planetary disc, which suggests they were formed by the same process, though one – Icarus –

1

1. Ida's 56-km (35-mile) length and a narrow 'waist' suggest that it was formed by two objects being rammed together.

2. An artist's view of Icarus glowing red hot at its closest approach to the Sun – 30 million km (19 million miles) closer than Mercury.

cuts in from below the disc, as if it were once ranging freely between the stars and was captured by the Sun.

Large and small meteorites

Meteorites are varied objects, characterized by their chemical composition and their history. They may be predominantly of stone, or iron, or both. All have varying amounts of oxygen, some have a high carbon content, and some even contain water, depending on how far from the Sun they originated. Some of those that are rich in nickel and iron – the same material as the cores of the inner planets – seem to have cooled very slowly, just one degree every million years. Originally, therefore, they must have been contained within a blanket of rock at

1

least 100 km (62 miles) across. These mini-planets were large enough to generate their own internal heat, and develop both a core and a crust. Over millennia, the crust was smashed by collisions with other bodies, leaving the naked core, several of which have proved useful as sources of iron to pre-industrial cultures.

Most meteorites are too dense to have derived from comets – they are most commonly the size of a pebble or small stone, but they can be larger, and occasionally very large indeed. Almost since the dawn of history, people have known of their celestial origin, and often venerated them as divine. In the Bible, the Ephesians worshipped the goddess Diana and 'the thing that fell from the sky', while the Muslim holy of holies, the Black Stone in the Grand Mosque, Mecca, is almost certainly a meteorite, blackened by its passage through the air.

1. Meteors from the Leonid meteor shower in November streak earthwards against a background of time-exposed stars.

2

2. This meteorite struck near Hoba West, Namibia, in prehistoric times. Found in 1920, it weighs 60 tonnes, one of the largest objects to survive its impact.

3. Found in the Atacama Desert, Chile, in the early 19th century, this stony-iron meteorite was forged during the evolution of the Solar System.

Since meteorites were traditionally regarded as magical, scientists dismissed them as objects of study until 1794, when a German physicist, Ernst Chladni, overcame the ridicule of his peers and argued convincingly that meteorites were actually bits of cosmic matter.

Every year, hundreds of meteorites plummet to Earth, with a flash and sometimes with a series of explosions. Although dull-looking things to untrained eyes, they have a great importance for astronomers because they are the only pieces of the Solar System that can be studied directly, other than Moon rocks brought back by US astronauts on the Apollo missions. Most fall in the sea, but several hundred are retrieved every year. One rich source is Antarctica, where glaciers sweep up meteorites and make them easily visible. Since the discovery of the first Antarctic meteorite in 1969, some 9000 finds have provided astonishing insights into the structures of their parent bodies.

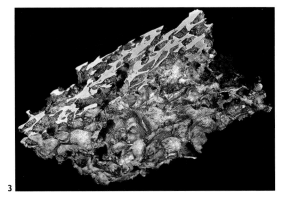

3

Larger meteorites, whether or not they are cometary remnants, have had a steady, and often drastic, influence on the evolution of the planets – the Earth included. Some that enter the Earth's atmosphere never make it to the surface. Between 1975 and 1992, American satellites recorded 136 explosions in the upper atmosphere, all probably small would-be meteorites. Other bodies survive

1

the fall, with dramatic results. Meteor Crater in Arizona, 182 m (600 ft) deep and 1200 m (4000 ft) across, was blasted out by a chunk of rock some 50,000 years ago. A 2000-kg (4500-lb) meteorite fell near Pueblito de Allende, Mexico, in 1969. In fact, it was from this meteorite that scientists were able to derive a good deal of information about the early Solar System, for it contains a peculiar form of magnesium, which could only have been produced

1. Magnified 50 times, this is a section of the 2-tonne meteorite that landed near Pueblito de Allende, Mexico, in 1969. Its complex structure contains 39 chemical elements.

2. Meteor Crater, near Flagstaff, Arizona, is 1.2 km ($^3/4$ mile) across and about 200 m (656 ft) deep. The 60-million-tonne asteroid that blasted it out vaporized on impact.

★ On 9 October 1992, a fireball streaked over New York state and landed in the town of Peekskill, smashing through the rear wing of a parked car.

2

within a massive stellar explosion. The magnesium must therefore have been blasted across interstellar space before being incorporated into the Allende meteorite, so here, in a sort of interplanetary fossil, was evidence of the event that could perhaps have triggered the formation of the Solar System.

The distinction between meteorites and asteroids is one only of scale. They are made of the same stuff, and originated in the same processes.

Asteroids, being larger, are rarer – and more dangerous should they strike a planet. Once or twice a century, something large hits the Earth, with a force that would be catastrophic if it struck a city. Every half million years or so, the Earth, like any other planet, can expect something much worse – all reminders that the Earth is still evolving in the same dynamic environment that shaped its evolution, and that of all its siblings.

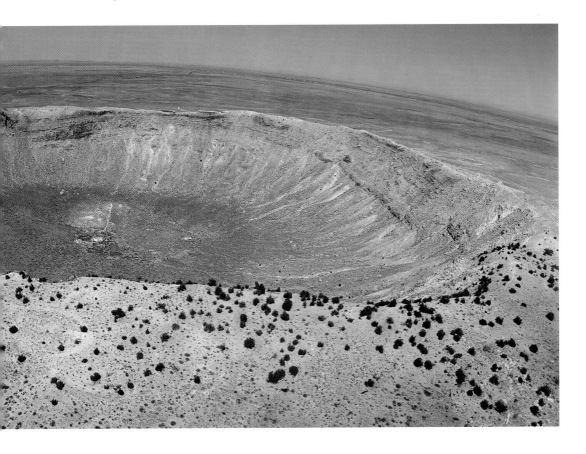

THE OUTER LIMITS

At one time, the arrival of individual comets and meteorites seemed to be isolated events, but in the early 19th century, evidence accumulated that allowed astronomers to see them, and other objects, as members of groups, all having their own origins and evolutionary histories. In the inner Solar System, detritus from the process of creation left a stony rubble, some of it safely corralled in dedicated zones, but some careering eccentrically between the planets. In the outer Solar System lie dormant comets by the billion, forming a diffuse and unseen halo around the distant Sun. This remote region divides into two domains – a nearer belt, the source of comets that orbit every few decades or centuries, and an outer shell so distant that a comet may take millions of years to make the round trip. Theory and evidence have transformed the Solar System into a place far more vast and complex than early astronomers could have dreamed.

Previous page: One of the most impressive visitors from the outer reaches of the Solar System last century was Comet Hale-Bopp, which illuminated the night skies in 1997.

DISCOVERY OF
THE ASTEROID BELT

A glance at a chart of the Solar System reveals two groups of planets: small inner ones and large outer ones, separated by a large gap. As Johannes Kepler, who worked out the distances of the planets in the early 17th century, suggested, it looks as if there is a planet missing. 'Between Mars and Jupiter,' wrote Kepler, 'I put a planet.'

Over a century later, Kepler's prediction of a planet between Mars and Jupiter was taken up by a little-known German astronomer named Johann Titius, who formulated a mathematical law that supposedly proved the existence of the missing planet. In 1772, the director of the Berlin Observatory, Johann Bode, popularized Titius's suggestion, which, as a result, is commonly known as Bode's Law. When Uranus was discovered in 1781, it fitted the 'law' to near-perfection, but

Neptune, discovered in 1846, does not. It seems Bode's Law is not a law after all; nevertheless, for almost a century it was virtually an article of scientific faith. The attempt to prove it turned out to reveal an unexpected and startling truth about the nature of our Solar System.

The search for the missing planet

Inspired by Bode's Law, Johann Schröter, Germany's foremost astronomer in the late 18th century, committed himself to finding the missing planet. Famous for his observations of the Moon, he ran an observatory in Lilienthal, near Bremen. In 1800, he invited five colleagues to organize the search. They

1. Most asteroids orbit in the Asteroid Belt – actually a dozen separate sub-belts – between Mars and Jupiter, but two small groups, known as Trojans, gather ahead and behind Jupiter in the same orbit.

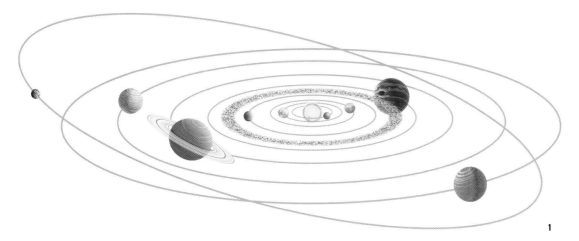

1

Discovery of asteroids more than 100 km (62 miles) in radius, 1800–1900

Date	Name	Discoverer
1801	Ceres	G. Piazzi
1802	Pallas	H. Olbers
1804	Juno	K. Harding
1807	Vesta	H. Olbers
1847	Iris	J. Hind
1847	Hygiea	A. de Gasparis
1850	Egeria	A. de Gasparis
1851	Eunomia	A. de Gasparis
1852	Psyche	A. de Gasparis
1854	Euphrosyne	J. Ferguson
1854	Amphitrite	A. Marth
1857	Doris	H. Goldschmidt
1857	Eugenia	H. Goldschmidt
1858	Europa	H. Goldschmidt
1861	Cybele	E. Tempel
1866	Sylvia	N. Pogson
1867	Aurora	J. Watson
1868	Camilla	N. Pogson
1872	Hermione	J. Watson
1892	Bamberga	J. Palisa
1896	Diotima	A. Charlois
1899	Patientia	A. Charlois

formed an association that soon acquired new members – 24 in all – nicknamed 'the celestial police'. Each promised to study a different part of the sky, checking all the stars of the zodiac to see if they could find an unknown moving object.

However, before the police could make much progress, they were beaten to it. In Palermo, Sicily, the Italian astronomer Giuseppe Piazzi was cataloguing stars when, on 1 January 1801, he spotted a moving point of light in the constellation of Taurus. He followed it for six weeks, but then – as he explained in a letter to the celestial police – he decided it was a tail-less comet, not a planet. In any case, by the time his letter arrived, the tiny moving dot had vanished.

Luckily, his records were good enough for the great German mathematician Karl Gauss to work out the object's orbit and predict its position, which he did so accurately that Heinrich Olbers, one of the original policemen, found it again almost a year later. This was no comet, but a new planet, just where it was supposed to be, in the gap between Mars and Jupiter. It was named Ceres, in honour of the patron goddess of Sicily.

Nothing but rubble

But there was something wrong: Ceres was so small – only 940 km (580 miles) across – that it could scarcely count as a planet. Perhaps there was another 'proper' planet? The search continued, and in March 1802, a few weeks after Ceres was named, Olbers found another moving dot, which was named Pallas. Olbers, an amateur who had set up

1. Karl Gauss was a mathematical genius who worked out the orbit of Ceres, the first asteroid to be discovered.

2. Vesta, the fourth asteroid to be discovered, is the third largest – 510 km (320 miles) across.

an observatory on the top of his house in Bremen, suggested that these two 'minor planets' could be the remains of a larger object that had broken up. If so, there could be more.

Lo and behold, in 1804 Karl Harding, Schröter's assistant, found a third (Juno), and Olbers a fourth (Vesta, in 1807). Instead of a single large planet, the police had found a collection of minor ones, or 'asteroids' (star-like objects) – as the English astronomer Sir William Herschel called them.

Was that all? The celestial police went on searching, without success, until, in 1815, they disbanded in the belief that they had found all the asteroids. In 1830, another amateur, Karl Hencke, took up the search. After 15 years, he found two more asteroids. Then, as techniques improved,

 BODE'S LAW

Johann Titius's 'law' predicted the existence of a planet between Mars and Jupiter. Popularized by Johann Bode (right), after whom it is named, it worked like this. Take the sequence 0, 3, 6, 12…and keep on doubling. Now add 4 to each number: 4, 7, 10, 16… The figures give the proportional distances of the known planets, starting with Mercury, with remarkable accuracy. But between Mars (16) and Jupiter (60), there is a gap (28) at roughly seven times the distance of Mercury from the Sun. This was where Titius and Bode predicted the existence of an undiscovered planet. As it turned out, the gap contains not one planet, but tens of thousands of 'minor planets' – the asteroids.

there came a rash of new finds: six more were discovered by 1850, and 432 by the end of the century, 92 of them found by one man, the French astronomer Auguste Charlois.

The American astronomer Daniel Kirkwood (1815–95) was the first to notice something odd about the distribution of the asteroids. Not only do they orbit in the path of the 'missing' planet, but their orbits are scattered. They form bands, with gaps between. This is because Jupiter's gravitational pull affects asteroids at crucial distances from the Sun, sweeping some orbital sections clean, forcing asteroids to follow others. The empty 'slots' are still named after their discoverer: the Kirkwood Gaps.

1. This compressed and idealized view from above Jupiter's moon Io portrays some of the Kirkwood Gaps in the Asteroid Belt.

2. Ceres, the largest asteroid at 940 km (580 miles) across, has not been photographed close up. This is a computerized view.

Now, there are some 10,000 named asteroids, from A'Hearn to Zyskin, and tens of thousands of numbered ones, with the population increasing by many thousands each year. Astronomers know that they are not, as the celestial police thought, the remains of a shattered planet, but the rubble of one that never formed, kept apart by the force exerted by their giant neighbour, Jupiter.

THE ASTEROIDS TODAY

Not long ago, even the larger asteroids were little more than enigmatic dots. Now, with the help of many types of telescope and interplanetary probes, we know a good deal more about them.

Ceres, the largest – it is almost twice as big as any other asteroid – lies in the middle of the belt and acts as a sort of frontier post. The asteroids closer to Earth are lighter in colour and resemble terrestrial rocks; those farther out in the colder

regions are darker because they are covered with soot-like carbon compounds. Ceres is one of these so-called carbonaceous asteroids, and also contains a large number of water molecules trapped in its minerals. At one time, Ceres must have been gathering smaller bodies of rock and been on its way to becoming a full-sized planet until its progress was stopped by the influence of Jupiter. For astronauts on the surface of Ceres – landing perhaps to replenish their supplies of minerals or even water – there would be little danger from the other free-floating bits of proto-planet, certainly nothing like the tumbling rocks beloved of science-fiction movies. Every few months, our astronauts might notice an asteroid drifting across the sky, but it would be no more than a bright, star-like object. Any colonists would have to wait many lifetimes for a collision.

Recent discoveries

During the 1990s asteroid science came of age as two probes examined four asteroids, with several other close encounters to come in the next decade. The first to be photographed was Gaspra, an ungainly angular monolith bypassed by the *Galileo* spacecraft on its way to Jupiter in 1991. Some 19 km (12 miles) long, it is about the size of Mount Everest. All newly revealed objects spring surprises on astronomers: Gaspra's surprise was that it possesses only small craters. Either it is quite a new asteroid, or it was recently blasted out of some larger body. The surprises of *Galileo*'s next subject, Ida, were that it has its own little moon, Dactyl, and

2

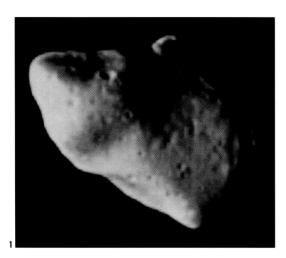

1

is covered with boulders, suggesting that it may prove exogeologists (those who study the geology of outer space) right – asteroids are large heaps of rubble.

When the *Near Earth Asteroid Rendezvous (NEAR)* spacecraft passed Mathilde in 1997, it revealed an object very different from Gaspra. Mathilde, 60 km (37 miles) across, has four massive craters. In fact, it looks like half an object.

NEAR went on to study Eros, chosen because it is one of those known as an 'Earth approacher' (other types threaten Earth more directly). Eros, the first asteroid to be given a male name, was discovered twice, almost simultaneously, by

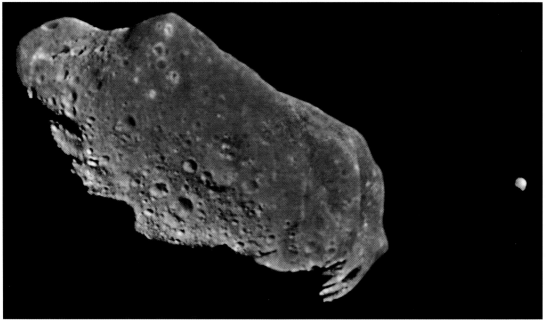

2

1. The tooth-like Gaspra, 18 km (11 miles) long, was the first asteroid to be encountered by a spacecraft (*Galileo*, in 1991).

2. Ida was the first asteroid to be found with a moon, Dactyl (to its right).

3. A painting shows the *NEAR* spacecraft in orbit around the 33-km (20-mile) Eros. In October 2000 it manoeuvred to 6 km (4 miles) above the surface.

3

Gustav Witt in Berlin and by the 19th-century's top asteroid hunter, Auguste Charlois, in Nice. Its relatively close approach makes it of interest because, from its mass and orbit, astronomers will be able to refine our knowledge of the Earth–Moon system. Only 33 km (20 miles) long, this orbiting mountain has a very weak gravitational field – a 45-kg (100-lb) object on Earth would weigh only 28 g (1 oz) on Eros – but it was enough to hold *NEAR* in a slow-moving orbit. After a year, its fuel ran low and NASA crash-landed it on Eros on 12 February 2001. Until then, it had revealed stunning views – impact craters large and small,

ridges, and small plains in varied colours, all material for a totally new branch of space science: asteroid geology.

Mixed identity

The remnants of our Solar System's origins are not limited to the Asteroid Belt: several of them swing in erratic orbits beyond Jupiter. The history of the first to be discovered also offers a warning that these objects are often not easy to categorize.

In 1977, the astronomer Charles Kowal noted something orbiting between Saturn and Uranus. At

first he thought it was a comet, but it lacked a tail, so he listed it as an asteroid. In honour of its dual nature, he named it Chiron, one of the centaurs, the creatures in Greek mythology that were both horse and man. Chiron (not to be confused with Pluto's moon Charon) was fair-sized for an asteroid, some 200 km (125 miles) across. So matters rested for 11 years. Then, as Chiron's orbit took it closer to the Sun, it suddenly doubled in brightness and developed a fuzzy aura of dust and gas, known as a coma. Chiron, the medium-sized asteroid, became a very large comet until it retreated into deep space again. One day, perhaps, its unstable orbit will bring it near the Earth and turn it into the brightest comet ever.

Chiron turned out to be one of a group of objects in elliptical orbits in the region of the giant outer planets. Seven are known, of which two others (Pholus and Nessus) are named after centaurs. When more of these objects are discovered, perhaps they, too, will reveal a dual nature and be named accordingly.

Another strange object with a mixed identity is Comet Schwassmann-Wachmann. Discovered by the two Germans after whom it was named in 1908, it is virtually unknown to the public because it is so hard to see, but it intrigues astronomers. It is a short-period comet, in an unusually circular orbit, making one circuit of the Sun every 15 years, approximately in the same orbit as Jupiter. Since it never approaches the Sun more closely, it remains more like an asteroid than a comet for most of the time. But every year or so it erupts, throwing off a cloud of gas and debris, and increasing its brightness 300-fold. Oddly, its tail takes on a spiral shape – apparently the nucleus, which seems to be about 40 km (25 miles) across and spinning, ejecting matter just as water is sprayed from a garden sprinkler.

Perhaps this comet is, as it were, on the brink of comet-hood. No one can yet explain its activity, but it might work like this: imagine a loose, porous surface of dark debris, which is left behind as the interior gases dissipate. Like the mess on a roadside snowdrift as the snow melts in spring, the surface forms a crust, trapping gases inside. When the pressure builds up, it bursts through the surface like a volcano. Then the exhausted nucleus returns to its former state, until the next explosion. One day, perhaps, when Schwassmann-Wachmann is better understood, it will be reclassified as a centaur.

The most distant object recorded from Earth is Asteroid 1996 TL66, whose orbit takes it out to 20,000 million km (12,500 million miles).

ON THE EDGE

Forty years ago, at the beginning of the Space Age, Pluto was routinely described as the most distant object in the Solar System. It seemed quite remote enough – 40 times farther from the Sun than the Earth, a distance that light itself takes five hours to cross. Surely no other objects could exist that far out, and anyway, if there were any, it seemed they would be too remote and small to be of any significance. But recent discoveries and theories suggest that if the Solar System is seen as a city, the planets occupy only its innermost heart. Beyond them, according to current models, lie two diffuse regions, one that reaches to twice the distance from the Sun to Pluto, and a second stretching to the very edge of the Sun's gravitational reach, an immense gulf 5000 times the distance to Pluto.

The Kuiper Belt

The first region is the source of short-period comets. This belt of icy bodies is named after the Dutch-American astronomer Gerard Kuiper, who suggested its existence in 1951. The suggestion became more attractive in 1980, when Julio Fernandez of the National Astronomical Observatory, Madrid, pointed out that most short-period comets orbited in the same plane as the planets, and ought therefore to have their origins in an extension of the disc of dust and gas from which the planets had condensed. For a few years, this remained a theory. Then, in the mid-1980s, the USA's *Infrared Astronomical Satellite (IRAS)*

1

2

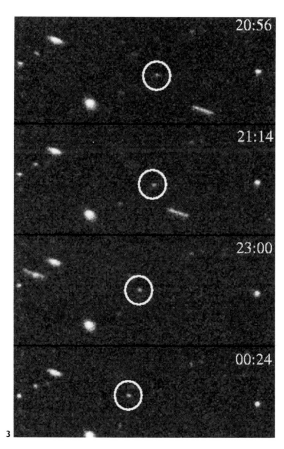

20:56

21:14

23:00

00:24

3

1. Pluto, with its moon Charon, was once believed to be the Sun's outermost object.

2. In 1983, *IRAS* found a Kuiper Belt around another star.

3. In 1992, the fuzzy spot (circled) moving slowly across the sky in a succession of shots was the first evidence of objects beyond Pluto. QB$_1$ became the first trans-Neptunian object.

photographed such a belt around a star in constellation Pictor. Then, in 1992, David Jewitt and Jane Luu at the University of Hawaii spotted a tiny object, about 320 km (200 miles) across, orbiting beyond Neptune – the first Kuiper Belt object.

At the time, Pluto's eccentric orbit had taken it inside that of Neptune for 20 years (it became the most distant planet again in March 1999). For this reason, the 1992 object, designated QB$_1$, was described as 'trans-Neptunian' rather than 'trans-Plutonian' – a wise decision, as it happens, because QB$_1$ is only just outside the orbit of Pluto, and in years to come Pluto will retreat beyond it. In fact, Pluto itself could be classified as a Kuiper Belt or trans-Neptunian object, and to add to the terminological confusion, some astronomers refer to all these bodies as 'Plutonians'.

Discoveries of trans-Neptunian objects came thick and fast: five in 1993, and increasing numbers every year since. At the time of writing, there are 343 of them, with the list being updated daily. The Kuiper Belt has become a reality.

Astronomers theorize that there should be up to 7000 million of these comet-asteroids, of which some 70,000 could be planetesimals up to 765 km (475 miles) across, with another 200 million measuring 10–20 km (6–12 miles) and the rest 1.6 km (1 mile) across, or less. Any one of these could be thrown from its orbit and become a live comet. In total, Kuiper Belt objects would not amount to more than a few per cent of the Earth's mass. Astronomers estimate that large bodies would be separated by about 162 million km (100 million miles) – the Earth–Sun distance.

1. In an artist's impression of the Oort Cloud, as seen from the Sun's nearest neighbours, the Sun's halo of dormant comets stretches out right to the triple-star system of Alpha Centauri. In fact, the Cloud would not be visible to the naked eye.

1

The Oort Cloud

The second cometary 'sink' is the source of long-period comets, those that take anything from 200 to more than a million years to orbit the Sun. This is the cometary cloud named after the Dutch astronomer Jan Oort, who suggested its existence in the 1940s after studying the orbits of 19 long-period comets.

This cloud, ranging from 6000 to 200,000 times the Earth–Sun distance, may contain at least 190,000 million, and perhaps as many as 10 million million, comets. Despite these immense numbers, the amount of material involved is remarkably slight – the whole cloud has an estimated mass 40 times that of the Earth, or one-tenth that of Jupiter, thinly spread across its vast domain.

The cloud is the result of a vastly extended and complex game of three-dimensional billiards. The young Solar System was filled with comets, which were gradually either swept up by the Sun and the planets or catapulted by gravitational interaction out of the Solar System entirely or into a sort of cometary graveyard, the Oort Cloud. As the cloud was built up by random ejections from the central Solar System, the objects would have been scattered in every direction. Unlike the Kuiper Belt, it would be a three-dimensional halo, encircling the Solar System like thistledown halfway to the nearest star – some two light years.

Oort showed that at these immense distances the dormant comets were held to the Sun by the minutest of gravitational forces and would be easily disturbed. What would disturb them?   ▷▷

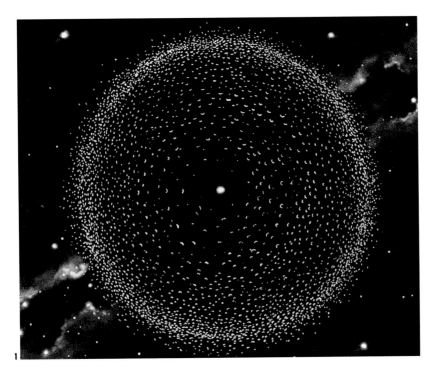

Previous page: Nemesis (right) is a theoretical companion to the Sun, invented to explain the regular pattern of extinctions on Earth. With a vastly elongated orbit, the faint 'death star' would return every 30 million years, destabilize the Oort Cloud and unleash a destructive swarm of comets.

1. In this artist's view, the Oort Cloud's mass of dormant comets circle the Sun. In the background is the Milky Way.

One answer is 'another star'. Stars are widely separated and seldom collide, but in their 200-million-year swing around the centre of the galaxy they steadily shift position. Every million years, a dozen or so stars pass relatively close to our Sun, not close enough to affect the Earth directly, but close enough to stir the Oort Cloud. It is, in Oort's words, a 'garden, gently raked by stellar perturbations'.

This scenario has added dimensions. If other stars have Oort Clouds – no one has seen our own, let alone that of another star – then a close approach could allow the two clouds to mingle. A third element of instability is the distant, steady pull of the centre of the galaxy, which would vary minutely across the cloud's four-light-year diameter. Finally, very rarely, once every 3–500 million years, the Sun bypasses one of the so-called 'giant molecular clouds' (or GMCs, to give them their acronym), the raw material of some future star. Together and separately, these influences would be enough to throw vast numbers of Oort Cloud objects into new orbits, showering the inner Solar System with comets, sending a new comet to startle Earth-bound eyes every few days rather than every few years. Every 100,000 years or so, comet watchers would see a dozen comets all at once.

Comet Hale-Bopp

The Oort Cloud may still be invisible, but astronomers recently saw a messenger from these distant regions. On a clear night in July 1995, Alan Hale was observing from his New Mexico home when, in a star cluster known as Messier 70, he saw a faint blob that should not have been there. Near Phoenix, Arizona, an amateur astronomer, Tom Bopp, happened to look at the same cluster at around the same time and saw the same glowing spot. Both realized that, by pure chance, they had seen a comet. Both reported their find simultaneously, so the comet was given both their names.

Two years later, Comet Hale-Bopp became one of the wonders of the northern hemisphere for its clarity, beauty and extraordinary qualities. Its speed and orbit showed that it had come from way beyond the Solar System, in fact from the Oort Cloud, but it had not been formed there. The Hubble Space Telescope revealed that the comet was voiding water at a massive rate – some 9 tonnes every second, a rate that actually increased as it rounded the Sun to a peak of 1000 tonnes of dust and 130 tonnes of water per second. Only its immense size – an estimated 40–80 km (25–50 miles) across – guaranteed it reserves enough for such an outpouring. It showed no trace of neon,

 PROBING THE DOMAIN OF THE COMETS

Four probes launched in the 1970s have now travelled into the Solar System's suburbs, where comets by the million circle as yet unseen. Travelling some 500–650 million km (300–400 million miles) a year, they took a decade to reach Pluto and the Kuiper Belt. Only one, *Pioneer 11*, has ceased to function. It died in January 1995, when it was 42 times as far from the Sun as the Earth – more than 6400 million km (4000 million miles). The other three – *Pioneer 10* and *Voyagers 1* and 2 (right) – are still transmitting more than 20 years after their departure. Having travelled some 55 times the Earth–Sun distance, they have entered the domain of the comets. Slight shifts in their path, caused by the gravitational pull of the dormant comets, may allow scientists to calculate their numbers. Eventually, all will drift on towards the Oort Cloud, where billions of comets reach halfway to the nearest star. If the *Voyagers* and *Pioneers* survive to enter the cloud, it will be as dead objects – artificial asteroids – in 2000 years' time. They could well continue their journey unscathed, finally leaving the Oort Cloud – and the Solar System – some 65,000 years later.

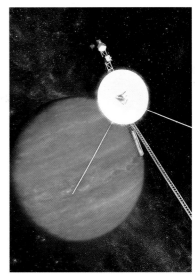

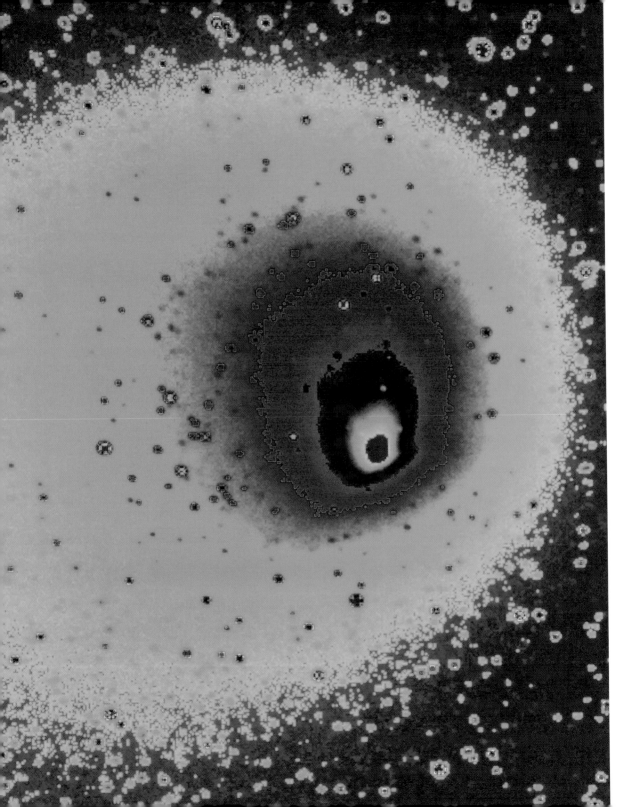

a gas that would have been present if it had formed in the frigid wastes of the Kuiper Belt. Possibly, Hale-Bopp was an early product of the relatively warmer parts of the Solar System, perhaps of the region of the gaseous giants. Some chance encounter with one of the giants accelerated it and slung it at an oblique angle clean out of the Solar System, into the Oort Cloud. There it drifted, dormant, for millions of years, until slowly it began to fall back in towards the Sun.

The journey took some 2000 years – Hale-Bopp's orbit suggests that it last flew past the Sun 4200 years ago. Only when it approached Jupiter did it burst into brief life, before vanishing again into the icy darkness.

2

1. (opposite) Comet Hale-Bopp was the most magnificent spectacle in the heavens that many people will see in their lifetime. Bursting into life in 1995 as it neared the orbit of Jupiter, it erupted in a complex coma.

2. Comet Hale-Bopp became visible from Earth as an indistinct blob.

3. For several months in 1997, Hale-Bopp's coma and tail dominated the night skies in the northern hemisphere.

3

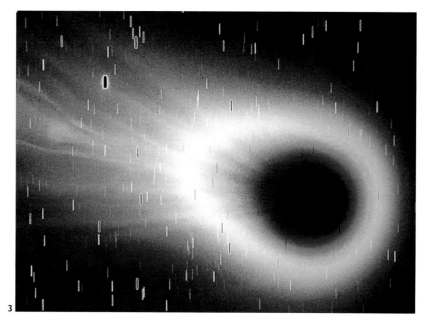

FROM MYTH
TO REALITY

FROM MYTH TO REALITY

Previous page: A sun-grazing comet on course for probable annihilation. Its tenuous tail flows away from the Sun under the pressure of radiation, while heavier particles form a meteor shower. The comet will either burn up or be torn apart, for few sun-grazers survive.

Throughout history, people have looked to the stars for guidance. Comets, the most puzzling and random of heavenly objects, seemed particularly ominous. Only in the 17th century, as prejudice and superstition gave way to science, did comets become comprehensible. Pride of place went to the best-known comet of all, Halley's, named after the man who worked out its orbit and predicted its return. It is still the only comet to have been photographed close up. Earth-based research carried out over two centuries, and deep-space probes over two decades, have now combined to describe a whole range of comets – bright and dim, steady and unstable, returnees and one-offs – and relate them to a wider family of asteroids, meteors and meteorites. All comets are ephemeral, destined to burn away to dust, decay into inert and invisible lumps of rock, or crash to fiery deaths on the Sun or one of the planets.

IN THE GRIP OF SUPERSTITION

In the ancient world, philosophers from Babylon to Rome agreed that comets meant bad luck. In the original Latin, a 'disaster' was something caused 'by a star' (*aster*). Some agreed with Aristotle that comets merely presaged winds, drought and cold weather. They were, he argued, exhalations emitted by the Earth, which ascended to the upper atmosphere and were ignited by the motions of the sphere that supposedly carried the Moon in its path. The same meteorological causes lay behind comets, winds, tidal waves and earthquakes – all horrors, but at least all natural phenomena.

Some argued for a deeper, more sinister connection between comets and catastrophes. Although Aristotle suggested that comets had a natural origin, their random appearance and equally inexplicable disappearance always kept open the idea that they were supernatural, not natural. Manilius, a 1st-century AD poet and astrologer, saw malign cometary influences behind the murder of Julius Caesar in 44 BC and the civil war that followed, a time of universal catastrophe – of incessant lightning, eclipses, eruptions, tidal waves, floods and earthquakes; of the births of monsters; of groaning graves; and of pathless forests echoing with disembodied voices.

No one doubted that comets were omens of some doom or other. A comet, declared Manilius's contemporary, the statesman and intellectual Seneca, 'is a sign of something that will happen'. The emperor Augustus tried to escape a fate seemingly presaged by comets in AD 9 and 11 by

1

1. In a French illustration of 1857, a comet goddess spreads destruction. The Greek word *kometes* means 'long-haired [star]'.

1. In the Bayeux Tapestry, the English quail at the comet – Halley's – that presages their defeat at French hands in the Battle of Hastings, 1066.

2. In his own atlas, Tycho Brahe presides over his theory of planetary orbits. His rigorous observations of the 1577 comet proved that comets were extra-terrestrial.

prohibiting diviners from calculating a person's time of death. It made no difference: his murder in AD 14 was heralded by a blood-red comet. Blighted crops, insurrection, civil war, the death of kings – all of these might be expected if a comet appeared, for both misfortune and heavenly unrest were linked by unseen causes.

A sign of God's wrath

It was a view that held sway for another 1600 years: Christianity inherited the fears and beliefs of its predecessors. Synesius of Cyrene, a contemporary of St Augustine in the 5th century, wrote that these evil stars, with their offensive hair, 'foretell public disasters, enslavements of nations, desolations of cities, the deaths of kings, nothing small or

moderate, but everything that exceeds disasters'. In the late 10th century, as the year 1000 approached, parts of France were convulsed with unrest, a near-revolution caused by a popular revulsion against land seizures by aristocrats. The appearance of a comet intensified the fear that the predictions in the Book of Revelation were about to be fulfilled: that the Devil, bound by Christ 1000 years before, was about to be let loose, and that Christ would return to smite him in a final apocalyptic battle. In the event, the comet vanished, the fears subsided and the year 1000 passed without great upset.

But 66 years later, another comet, recorded by English monks and French weavers, seemed to mirror the Norman invasion. Queen Matilda, wife of William the Conqueror, commemorated her husband's victory at the Battle of Hastings by commissioning a tapestry in which awestruck Englishmen point skywards and whisper doom into King Harold's ear, while ghostly Norman ships foreshadow his defeat. In 1314, the French king Philip IV died after a fall. It was caused by a wild boar tripping his horse, but an account of his death illustrates the 'real' cause – a comet.

Repeatedly in the 16th century, when Europe was convulsed by the ferment unleashed by Protestant reformer Martin Luther, churchmen saw war, pestilence, revolution and famine presaged by comets. 'Whatever moves in the heavens in an unusual way,' wrote Luther, 'is surely a sign of God's wrath.'

A whole pseudo-science of cometary astrology arose. The great Danish astronomer Tycho Brahe thought that because the comet of 1577 first appeared 'with the setting of the sun', it heralded calamities west of Denmark. The English astrologer William Lilly assured his gullible countrymen that since the comet of 1678 appeared in the constellation of Taurus, it would therefore affect Russia, Poland, Sweden, Norway, Sicily, Algiers, Lorraine and Rome, but luckily not England.

Comets readily became political: the turbulent times of the English Civil War (1642–51) and the restoration of the monarchy in 1660 proved particularly fertile ground for astrologers. To pamphleteers, writing with all the wisdom of hindsight, the comets of 1664 and 1665 heralded the Plague, the Great Fire of London and the Anglo-Dutch War, while one in 1677 ushered in the Popish Plot – when Catholics planned to kill Charles II and burn London. The obvious conclusion drawn by propagandists was that Catholicism was doomed.

⭐ The earliest recorded account of a meteor shower was in China in 1809 BC, when meteors were said to 'fall like a shower at midnight'.

THE COMING OF SCIENCE

It was Tycho Brahe himself, despite being a believer in the dire effects of comets, who opened the way to scientific progress. In the 16th century, theory was dominated by the ancient belief that the planets were carried round on fixed 'spheres', of which the outer ones were reserved for the stars; that the Universe was perfect and therefore unchanging; and that comets, since they moved so wildly and were so changeable, must therefore be atmospheric phenomena. Tycho, an arrogant nobleman of independent means, was able to fund researches that made him the best observer of his day. His breakthrough, and one of his great claims to fame, came in 1572, when he recorded the appearance of

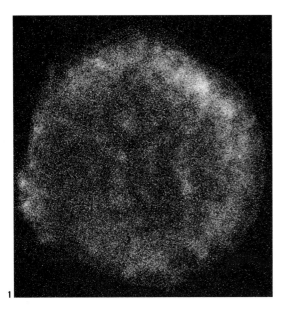

a new star, which we now know was an exploding star, or supernova. He was able to show that it was a very distant object, not, as Aristotle would have claimed, a product of the atmosphere, thus proving that the heavens were not immutable after all.

Tycho came to similar conclusions in his account of the great comet of 1577. As he wrote in his book on the subject, the orbit of the comet was at least three times the distance of the Moon. And since it approached the Earth and retreated again, it could not be carried on any fixed sphere. His evidence and arguments formed part of the new view of the Solar System that evolved in the 16th and 17th centuries.

Newton's laws

But the nature of comets remained a mystery, and scientists and lay people alike continued to regard them as omens. It was not until 1682 that another comet heralded the onset of a more common-sense approach, thanks to the English astronomer Edmond Halley. Fascinated from boyhood by mathematics and astronomy, Halley had made a name for himself in his early twenties when he published a star catalogue. He had been made financially independent by his father, luckily for posterity, because it was he who persuaded Isaac Newton to publish his theory of gravity, and paid for the publication of Newton's monumental *Principia Mathematica* in 1687. Three years previously, Newton had proposed the first unified theory of planetary motions, suggesting that comets and planets were governed by the same laws. The *Principia* stated

PHILOSOPHIÆ

NATURALIS

PRINCIPIA

MATHEMATICA.

Autore *JS. NEWTON*, Trin. Coll. Cantab. Soc. Matheseos Professore Lucasiano, & Societatis Regalis Sodali.

IMPRIMATUR.
S. P E P Y S, Reg. Soc. P R Æ S E S.
Julii 5. 1686.

LONDINI,

Jussu *Societatis Regiæ* ac Typis *Josephi Streater*. Prostat apud plures Bibliopolas. *Anno* MDCLXXXVII.

2

1. Tycho's Star, which he discovered in 1572, is now known to be a supernova remnant, the remains of a star that exploded about 10,000 years ago.

2. Edmond Halley (above) not only identified the comet named after him but funded Isaac Newton's *Principia Mathematica* (top right), which stated the laws of gravitation – proof of which was provided by Halley's Comet.

these laws in rigorously mathematical terms that applied to all matter, including comets. Far from being as wildly eccentric as their orbits, comets could turn out to be the ultimate proof of the Universe's underlying regularity. No more would astrologers and pamphleteers be able to present comets as harbingers of doom. Order would return to both heaven and Earth.

In 1695, Halley took up the challenge. He checked through records of past comets and used them to calculate their orbits. There was not much to go on because most observers had assumed that comets were meteorological and had not bothered to record their progress against the stars. Besides, it was extremely hard to tell the difference between an 'open' parabolic orbit, which would mean that a comet was a one-off, as ancient beliefs implied, and an elliptical one, which would eventually bring it back. At last, some records started to make sense.

Three sets of observations – of comets that appeared in 1531, 1607 and 1682 – were virtually identical. From this, as the Royal Society noted, '[Halley] concluded that it was highly probable, not to say demonstrative, that these were but one and the same comet, having a period of about 75 years'. The conclusion was clear, as he stated in his *Synopsis of Comets* (1705): '…I dare venture to foretell, That it will return again in the Year 1758.'

Halley's research was the first prediction based on Newton's laws. Halley himself died in 1742, aged 86, but as the predicted date of the return approached, 'comet fever' spread. John Wesley, the Methodist preacher, said it would set the Earth on fire. In America, John Winthrop, professor of mathematics and natural philosophy at Harvard,

agreed with the apocalyptic view that a comet's tail could drench a planet in enough water to cause another Great Flood. Comets were 'God's tools' – who knew what He might choose to do with them?

Astronomers were excited for other reasons. The records showed small variations in the orbit of Halley's Comet. Its period varied by a year or more, according to which giant planets affected its progress. Depending on its return date, therefore, Halley's Comet would either vindicate or undermine Newton. Feverish calculations by the French mathematician Alexis Clairaut predicted the comet would pass the Sun in the middle of April 1759. Almost perfectly on schedule, a German amateur named Georg Palitzsch spotted the comet on Christmas Day 1758. It passed the Sun on 13

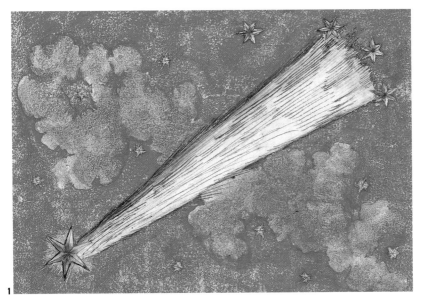

1. From records of its orbit, Halley was able to show that the comet of 1531, here portrayed in a 16th-century view, was the same one he saw in 1682.

2. Halley's Comet, seen by Giotto in 1301, was transformed by him into the Star of Bethlehem in his *Adoration of the Magi* (1304) – a rare use of an event more often seen as presaging doom than salvation.

March. Calculations drawn from observations taken over a period of 150 years had led to a prediction accurate to within a month – a brilliant vindication of Newton's laws.

After that, it was simple to calculate when Halley's Comet would appear again, and when it had appeared in the past. Working backwards, past appearances included one in 1456, which struck fear into Muslims besieging Belgrade, and one in 1301, when the Italian painter Giotto included it as the Star of Bethlehem in his *Adoration of the Magi*. Indeed, the last 30 visits all seem to have been noted, the earliest sighting being one made in China in 240 BC. It put in a significant appearance in 1066, as the Bayeux Tapestry records – not a harbinger of Harold's defeat at all, but coincidental proof that the heavens proceed according to their own Newtonian laws with no regard for battles.

HALLEY IN CLOSE-UP

After Halley, the next of the great comet-hunters
was the French astronomer Charles Messier. One of
his claims to fame was that he drew up a catalogue
of 103 nebulas, or star-clouds, many of which are
now recognized as sister galaxies to our own. In
their official designations they still bear his name,
usually reduced to its initial – our nearest galactic
neighbour, the Great Spiral in Andromeda, is
Messier (M) 31. But his main passion was comets:

nicknamed the 'ferret of comets' by Louis XV, he
discovered 21 of them.

Messier's mantle was inherited by his
countryman Jean Louis Pons, who started his
professional life as caretaker in the Marseilles
Observatory and worked his way steadily upwards
to become an observatory director, ending his
career in Florence's Museum Observatory. Pons
discovered 36 or 37 comets (sources vary), more
than anyone else in history. One of them was
named in 1818 after the man who computed its

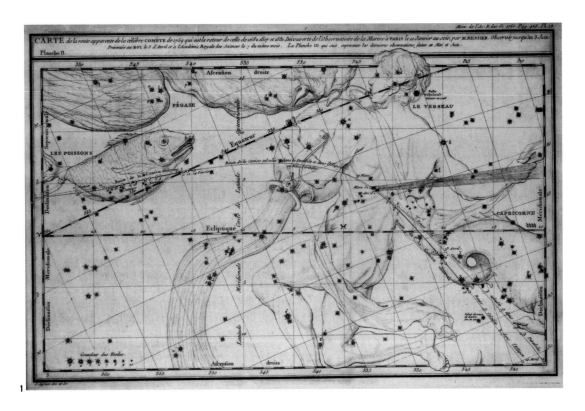

1

1. The 'ferret of comets', Charles Messier made this record of the track of Halley's Comet through the constellations Pisces and Virgo in early 1759.

2. On its return in 1910, Halley's Comet put on a brilliant display, captured by this photograph taken at Helwan, Egypt.

2

orbit, Johann Encke of Göttingen. Encke's Comet was of great interest because it completed its orbit round the Sun in only 3.3 years, making it one of the closest of the major comets. Tracking backwards, Pons was amazed to discover that he had seen the same comet in 1805, before he knew its period.

Vanishing comets

Over the next century and a half, hundreds of comets were recorded, up to a dozen in some years. Astronomers agreed on their general characteristics – small mass, insubstantial nucleus, and an even more insubstantial tail that was a stream of gas and dust burned off by the Sun, like the 'smoke from a chimney' as one 19th-century historian of science

described it. If that was so, then comets would eventually burn out.

This theory was proved correct in a dramatic fashion by the fate of a comet named in 1826 after its discoverer, an Austrian army officer called Wilhelm von Biela. (In fact, it had been seen many times before, probably as far back as 1772, but no one knew it was the same one returning again and again.) Once its period was known – 6.75 years – astronomers tracked it carefully. In 1845, it astounded watchers by dividing in two. The twins reappeared on schedule in 1852. In 1858, viewing was poor, but astronomers eagerly looked forward to the double comet's return in 1866, when it was due to fly close to the Earth. What they saw was absolutely nothing. The comets had vanished. In 1872, again no comet appeared, but there was a

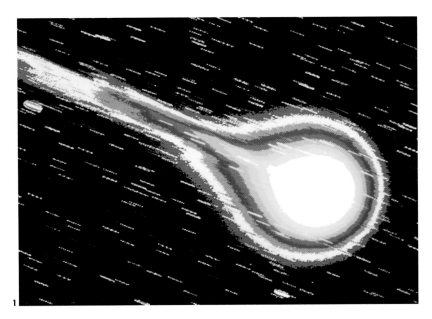

1. A computer-generated, colour-coded view of Halley's Comet shows its hot core burning off gas and dust.

2. The 16-km (10-mile) core of Halley's Comet as seen by the *Giotto* spacecraft in March 1986.

3. A false-colour image of Halley's Comet taken from a rocket reveals a cloud of neutral hydrogen, tens of millions of kilometres across.

spectacular meteor shower, with meteors filling the skies of Europe like fireworks, falling too fast to be counted. The display was clearly caused by the detritus from Biela's Comet, proof that meteors are mostly the dust from the trails left by comets.

By 1900, it was clear that the tail was not like the tail of a rocket. It is the radiant energy from the Sun that strips away a tenuous spray of dust from the head and scatters it 'downwind'. For this reason, as the comet swings round the Sun and heads off again into the outer reaches of the Solar System, its 'tail' precedes it, pushed by the solar wind, until it is so far from the Sun that all activity ceases and it becomes dormant until its return.

There remained many mysteries. What are comets made of? How exactly are the tails made?

Where do comets come from? What becomes of them? If they come close enough for the Earth to pass through their tails, do they ever hit the Earth? And what would be the effect?

A 'dirty snowball'

It is Halley's Comet that holds the key to much of what is known about these interplanetary wanderers because no other has been studied in such detail. For astronomers, it has several advantages: it is large, well placed for Earth-based observation, spectacular and, above all, regular. Moreover, it is unusual in having a 'retrograde' orbit – it loops round the Sun in the opposite direction to the Earth and other planets. Its return in 1986 offered a superb opportunity for close-up examination, undertaken by a small armada of spacecraft from Japan, the Soviet Union and Europe (the US Congress refused to fund an American mission).

The European Space Agency's *Giotto* spacecraft, named after the medieval Italian artist who recorded the comet, swept in to within 600 km (400 miles) of the nucleus. Astronomers had guessed that the core would be roundish, bright and about 5 km (3 miles) across. To their astonishment, *Giotto* revealed an irregular, coal-black, peanut-shaped object measuring about 8 × 16 km (5 × 10 miles), with a mountain and several craters, and with jets of gas bursting from a number of vents. The gas came from ice that lay beneath the surface, the surface itself being a thin crust of ash-like material, in effect carbon. Whipple would have been delighted: this

was a very dirty snowball (▷ p. 18). The comet was tumbling in a complex motion over a period of about five days, rotating and somersaulting at the same time, exposing itself to the heat of the Sun like a sausage on an automatic grill. It is the combination of rotation and heat, which increases as it nears the Sun, that brings it alive. As vents turn into the shadows, the carbon crust heals and seals them shut, while others on the sunny side explode as sub-surface gases heat and expand. The gas, and the dust blown away by the eruptions, produce the comet's surrounding cloud and its long glowing tail. The craft also discovered that Halley's tail included some surprisingly complex chemicals, such as sulphur and carbon compounds, which are basic constituents of life – proof that the Earth shares a common origin with comets.

2

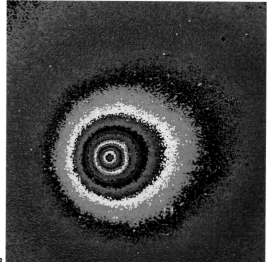

3

Halley's path

This mass of information provides Halley's Comet with a life history. Unlike most comets, which seldom cross Earth's orbit, Halley's Comet approaches to 87 million km (54 million miles), between the orbits of Venus and Mercury. Its close approach subjects it to more heat than most, and a greater loss of material. With every pass, it loses about 100 million tonnes, the equivalent of a layer 1.8–2.0 m (6–7 ft) thick. It only sustains this loss for little more than a year during its 76-year orbit, but it cannot last for ever. It has already been orbiting the Sun for 175,000 years, making 2300 passes, which

have reduced it from its original 32-km (20-mile) radius to half that size. It is about halfway through its life. If it survives without sustaining an impact or being flung out of its orbit by a close encounter with a major planet, it will last for another 2500 passes, another 187,000 years. What then? Either it will become an inert lump of rock, or it will evaporate.

What of the stream of dust and gas left behind by Halley's Comet? Surely the fallout from such a significant visitor should cause a dramatic display for Earth-based meteor-watchers? Oddly, it does not. It took over a century to understand why. The answers explain much about the path of comets and how they interact with the Earth.

1. (opposite) An imaginary view from a comet as it approaches Earth. Heated by the Sun, it sputters gases through its icy, carbon-rich surface.

2. Halley's Comet orbits in the opposite direction to Earth. The dates relate to its orbit in 1910–2010.

3. An Earth-based shot of Halley's Comet, time-exposed against a background of fixed stars, shows how hard it is to analyse a comet with an optical telescope.

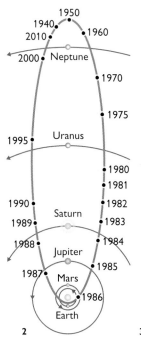

2

3

The explanation, which links the comet with two meteor showers, took 120 years to emerge. The story started in 1863, when an American astronomer, Hubert Newton, suggested on the basis of ancient records that a clearly defined meteor shower should occur in late April or early May. He was proved right seven years later. Observations revealed that the shower seemed to originate in a region of the sky around a star (η 61544) in the constellation of Aquarius. (The symbol η is the Greek letter eta.) Meanwhile, Alexander Herschel, the grandson of the great astronomer Sir William Herschel, had pinpointed another meteor shower, visible in October, that seemed to radiate from the constellation Orion. He did not suggest Halley's Comet as a source of the Orionids, but he did think the comet caused the Eta Aquarids, the meteor shower around η 61544 – a theory confirmed in 1886. Only in 1911 did it begin to seem likely that both streams were actually one. But it took more sophisticated techniques to prove it and explain why Halley's Comet does not make more of a show in Earth's atmosphere.

First, astronomers discovered that the particles from the two showers zip into the atmosphere faster than almost any other set of meteors – at 65 km (40 miles) per second. That fitted – Halley's retrograde orbit sends dust particles to meet the Earth head on. But why were they so faint? The answer became clear in 1986, when the two showers remained faint, despite Halley's flyby. Clearly, the Earth did not go exactly through its

1

track. So what is the relationship between comet, tail and meteor showers? The answer is that cometary tails are insubstantial things, made of particles all of which are in their own orbits, all subject to pressure from the solar wind. They drift and disperse, like gossamer. As Halley's Comet swings round the Sun, its tail does not directly intersect with the Earth's orbit. Over the centuries, however, it gradually diffuses – and only then do its fuzzy edges overlap the Earth's orbital path in two places, or 'nodes', and then some of its particles flash into view as Eta Aquarids (in April–May) or Orionids (in October). The dust that forms them was ejected centuries ago, and it will be many more centuries before its 1986 flyby contributes to meteor showers on Earth. And those showers will still be lighting Earth's skies for centuries after Halley itself has vanished.

1. A fine view of Halley's Comet with its 'mature' tail, taken from Siding Spring, Australia.

2. An early view of Halley's Comet shows that its tail has not yet been fully formed by the Sun's heat.

▷ COMETS AND THE ORIGINS OF LIFE

In the 1960s, astronomers spotted traces of a range of chemicals created by reactions in interplanetary and interstellar space. The 40 or so molecules included carbon, ammonia, formaldehyde and formic acid. In addition, amino acids have been found in meteorites. British astronomer Sir Fred Hoyle suggested that comets might have carried the building blocks of life to Earth and kick-started evolution. This idea struck many scientists as eccentric, but the *Giotto* space probe (right) found complex organic substances on the nucleus of Halley's Comet. And a NASA team has subjected comet-like gases to temperatures of about 8000°C to simulate the heat generated by a comet striking the Earth's atmosphere. Some molecules survived, and even formed new compounds, including water, carbon dioxide, methane, nitrogen and hydrogen sulphide. If more evidence accumulates, Hoyle's theory may join the mainstream.

OF TAILS AND METEOR SHOWERS

To date, some 1000 comets have been recorded. Compared to the billions roaming undetected, this is a small sample, but it is enough for astronomers to provide a unified view of comets, explaining all the major elements – their origins in the Oort Cloud and Kuiper Belt, their orbits, their life cycles and their fallout as drifting particles and meteors.

On astronomical timescales, they are all transitory objects, once they settle into solar orbit. They melt, erode, evaporate and finally vanish in a variety of related processes. Some blast out their gaseous tails until their fuel is exhausted and then sink into somnolence as asteroids. Some blow themselves to bits. Some are weakened and then torn apart by the gravitational pull of the Sun or one of the planets. In most, eruption is a steady process, at least while they are travelling through the domain of the planets. It increases as they approach the Sun, decreasing again as they depart. But some occasionally emit flashes, like sputtering volcanoes that erupt without warning, increasing their brightness many thousand-fold. No one is quite sure yet why some comets flare (as Comet Schwassmann-Wachmann does, ▷ p. 39), but one cause could be a collision with other small interplanetary bodies. After all, many asteroids carry impact scars and Halley's Comet was shown to have several craters. If an impact does not destroy a comet, it might well kick-start an eruption.

Not one, but two tails

The tails that jet from a comet's surface are complex things. The eruptions produce a faint aura of dust and gas, known as a coma. As astronomers had theorized and the *Giotto* space probe to Halley's

1. A false-colour profile of Comet Bennett in 1970 reveals the elements in its structure – nucleus, coma, dust-tail, plasma-tail – in a five-part scale of temperature.

1

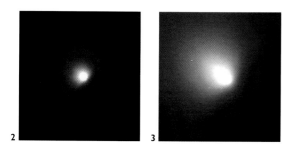

2

3

4

2,3,4. In 2000, the dim and distant Comet Linear provided a sudden display of instability by blowing off a piece of its crust. The pictures were taken over three days, 5–7 July.

establish the orbits of the parent streams and, if possible, work out their histories. A favourite in Europe is the Perseid shower, so named because it seems to originate in the constellation Perseus. Every August, the Perseids flash through the warm skies, to the delight of holiday-makers. Like comets, the Perseids have been tracked back in time through ancient records. We now know that the Perseids were active as long ago as AD 36.

The astronomers' puzzle

Of the 21 displays during the course of the year, a couple are still mysteries. One, the Quadrantid shower, first came to the attention of astronomers in January 1825, when it was recorded by an Italian astronomer, Antonio Brucalassi, and named ▷▷

Comet confirmed, solar radiation and the flow of charged particles that make up the solar wind push the dust out of the comet's feeble gravitational field and spread it 'downwind' into a dust tail that curves away from the Sun. But there is also another type of tail, which emerges close to the Sun. Intense solar radiation strips electrons from some of the gases, turning them into ions, forming a plasma that is responsive to the Sun's magnetic field. The result is a second tail, a plasma, which glows with an eerie blue light as the ionized atoms fluoresce in ultraviolet light. This tail also points away from the Sun, but it points directly outwards.

From the tails, of course, come the dust particles that form meteors. Meteor showers now have their own specialist studies, which involve research to

The greatest recorded meteor shower was the Leonid 'storm' seen in the USA and Canada on 12 November 1833. Meteors fell at an estimated rate of 30,000 an hour for seven hours.

Previous page: To an astronaut, a comet with its tails of dust and plasma would look like this as it passes the Earth and Moon.

1. In 1996, Comet Hyakutake put on a dramatic display with a relatively close approach to Earth – some 15 million km (10 million miles).

2. In 1965, Ikeya-Seki was the first comet to have its heat measured by infrared telescope.

1

after a defunct constellation, Quadrans Muralis (the Mural Quadrant), which was removed from the over-cluttered skies when today's 88 official constellations were adopted in 1922. Detailed records started only in the 1860s because the Quadrantids are something of a challenge. Visible only in the northern hemisphere, they peak after midnight on 3–4 January, falling at the rate of 100–190 per hour. They are faint and their point of origin is ill-defined. For any astronomer braving an icy night, it is like counting raindrops in a gale. The parent stream is apparently old and diffuse, and there is no trace of the original comet. Astronomers speculate that it broke up about 1500 years ago, leaving behind both the Quadrantids and a second puzzle, the Delta Aquarids, which arrive in July from a 'radiant point' around the star designated

δ (delta) 61540 in the constellation of Aquarius. If this is so, both streams foreshadow the fate of those comets that survive in orbit until their eventual extinction. All that is left is a scattering of dust, and then finally nothing at all. In another three or four centuries, the Quadrantids will have vanished.

Visible and invisible comets

Seen from Earth, every comet has its own individuality. Three or four times a century, there are comets so bright that they are visible to the naked eye even at midday. One of these appeared in 1910, when it was spotted in the pre-dawn sky over Johannesburg on 12 January. The following week, the Great Daylight Comet was seen across northern Europe. The most recent daylight comet was Comet

West in 1976. It will not be returning: about a week after its closest approach, it broke into four separate pieces.

Although comets are considered rarities by Earth-bound watchers, the ones that are visible are vastly exceeded by the ones that are never seen because their light is obliterated by the glare of the Sun. Only special circumstances and specialized techniques allow them to be visible at all. In 1948, an eclipse seen from Nairobi revealed the presence of a comet on the edge of the Sun's disc. That was the first glimpse of an object that a week later drew far enough from the Sun to be visible to the naked eye. It was nicknamed 'the Eclipse Comet'. Three other comets were seen to approach the Sun in the 19th century, inspiring a German astronomer,

A checklist of daylight and eclipse comets

1843: Unnamed February comet
1882: Comet Tewfik, during May eclipse
1882: September comet
1910: Great Daylight Comet
1947: Comet Rondanina-Bester (C/1947 F1), during May eclipse
1948: Eclipse Comet (C/1948 VI), during East African eclipse
1965: Comet Ikeya-Seki
1976: Comet West (C/1975 VI)
1997: Comet Hale-Bopp, during March eclipse

Heinrich Kreutz, to suggest that they were the remnants of a much larger comet that broke up.

Recently, satellites equipped with cameras designed to blot out the solar disc have revealed a whole new subset of such comets, which either zip around the Sun at a distance of only 8 million km (5 million miles) or less, or crash right into it. Some 75 sun-grazers have been discovered. They are all small and few survive the experience, either breaking up under the intense strains imposed by gravity and heat, or vaporizing entirely. It seems an odd coincidence that so many should approach so close – unless it is not a coincidence at all. Perhaps, as Kreutz suggested, they are all one family, tiny remnants of a huge mother-comet torn apart by successive passes close to the Sun. The truth about the Kreutz Group has yet to be established.

2

TARGET
EARTH

TARGET EARTH

As blockbuster movies and serious astronomers regularly remind us, we live in a shooting gallery. In the long term, we are at the mercy of wandering asteroids and comets. But the idea that impacts might have affected us directly – or may do so in the not-too-distant future – was dismissed as mere fantasy until recently. Now ancient beliefs that comets were heralds of catastrophe seem to be statements of fact, not superstition. New research suggests that comets may have had a crucial influence on human history on at least five occasions, while new asteroids are being discovered daily, many of them on orbits that cross that of the Earth. Catastrophic scenarios predicting the end of humanity, or even the world, may be more the stuff of fiction than immediate reality, but some lesser disaster is certain eventually – a discovery that focuses the minds of researchers and policy-makers alike.

Previous page: An artist's impression of a comet, its icy surface seething with violent eruptions brought on by the Sun's heat, approaching the Earth.

THE GROWING WEIGHT OF EVIDENCE

Early on the morning of 30 June 1908, a Russian peasant was taking a meal break, sitting beside his plough on the banks of the Angara River in central Siberia, when he heard a noise. 'I heard sudden bangs, as if from gunfire,' he recalled later. 'My horse fell to its knees. From the north side above the forest a flame shot up. Then I saw that the fir forest had been bent over by the wind, and I thought of a hurricane. I seized hold of my plough with both hands so that it would not be carried away. The wind was so strong it carried soil from the surface of the ground, and then the hurricane drove a wall of water up the Angara.'

The explosion, which occurred over the Tunguska River 200 km (125 miles) north of where the ploughman was sitting, was picked up in meteorological stations in St Petersburg, Berlin, Potsdam and London, where a second change in pressure was noted a day later: the echoes of the Tunguska Event, as it has been called, had travelled right round the world in the other direction. Travel was difficult in this remote area, and it took almost 20 years for scientists to investigate. When, in 1927, they went to the site of the explosion, they found a scene of devastation. There was no crater, but in a circle some 32 km (20 miles) across, trees lay felled, all splayed outwards from the centre of the blast.

For years, the impact remained a mystery. One obvious possibility was that the object was a

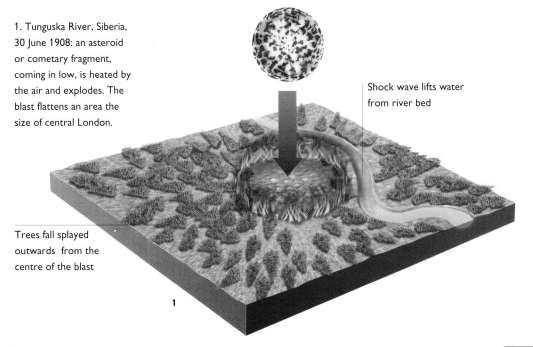

1. Tunguska River, Siberia, 30 June 1908: an asteroid or cometary fragment, coming in low, is heated by the air and explodes. The blast flattens an area the size of central London.

Shock wave lifts water from river bed

Trees fall splayed outwards from the centre of the blast

1

meteorite. But why the absence of a crater? Why would it have exploded high in the atmosphere? For years, the absence of hard facts inspired wild speculation, including one suggestion that an alien spacecraft had exploded. Now scientists are sure that the event was caused by an asteroid 50–60 m (165–200 ft) across, weighing some 100,000 tonnes and travelling at about 30 km (20 miles) per second on a very flat trajectory – a few miles in another direction, and it would have missed the Earth completely. As it was, its passage through the thickening air heated it so fast that the heat could not escape. The asteroid exploded some 10 km (6 miles) up and released a burst of energy estimated at 15 megatonnes – 15 million tonnes of TNT, or about 1000 times larger than the bomb that flattened Hiroshima in 1945. If it had exploded over a major population centre, it would have killed millions of people.

Catastrophism *v.* uniformitarianism

The explosion was in such a remote region that its immediate impact on scientific thinking was virtually nil. At the time, astronomers and earth scientists held the comforting belief that the Earth and its life forms had evolved steadily and majestically, from simple to complex, from primitive to advanced. It was a belief derived from two revolutionary views that had become part of the conventional wisdom: Newton's theory of gravitation and Darwin's theory of evolution. Although separated by three centuries, both were fundamental to the reassuring notion that the

Universe and the evolution of life were governed only by orderly processes. Catastrophes seemed reminiscent of pre-scientific superstitions, like the Flood and miracles. 'Catastrophism' was out; 'uniformitarianism' was in.

From the middle of the 20th century onwards, catastrophism made a modified comeback, as evidence emerged from diverse specialities, in particular the earth sciences and astronomy. In the earth sciences, palaeontologists were puzzled by two long-standing problems. A good place to study the first is in the Karoo Desert, South Africa's veldt of scrub, grass, rock and sand. Here, where springbok dart away from cars travelling the lonely route from Cape Town to Johannesburg, Earth's history from about 400 million years ago is written in sand-stones and shales originally deposited on the seafloor and thrust up at odd angles. Impressions of horsetails, mosses and ferns show how the first plants colonized the bare landscape. Then, in higher strata, appear a wonderful range of

1. The prophet of a new faith, Charles Darwin asserted that the key to the origins of mankind lay in the forces of natural selection working regularly over aeons. His theory made no allowance for catastrophic events.

2. Catastrophes, such as the Flood – shown in this view by the German pre-Raphaelite Julius Schnorr von Carolsfeld – remained an article of faith to the religious throughout the 19th century. As science undermined faith, catastrophes fell from fashion. Now they are back on the agenda.

2

amphibians and proto-mammals. Some 270 million years ago, at the start of the Permian period, species competed to exploit newly opened niches. Herbivores and their predators proliferated in size and shape, in an evolutionary dance that modified size, fleetness, teeth, defences, eyesight and claws for some 20 million years. Then, about 250 million years ago, the dance music came to a stop. It seems that almost half the area's species suffered near-simultaneous extinction. In a coda lasting some 5 million years, new species evolved to replace them. But finally came a universal silence, a great dying: the whole era that geologists call the

The Book of Revelation seems to record a close cometary encounter: 'There fell a star from Heaven…and many men died of the waters because they were made bitter.'

Palaeozoic ('ancient life'), the 300-million-year progression from small marine organisms to proto-mammals, ended. At sea, some 90 per cent of all invertebrate families vanished, including almost all corals, brachiopods and sponges. The trilobites, which for millions of years were the oceans' universal scavengers, vanished completely. Most families of fish, snails and clams disappeared as well.

The rebirth that then followed produced creatures of such size and variety that they have dominated the human imagination since they were first identified in the early 19th century. The dinosaurs and their relatives spun off subforms, as the mammals were later to do, colonizing the air and the sea for 140 million years. On land, the flowering plants emerged – evolving into more than 80 per cent of all plants – and insects proliferated. In the seas, ammonites – free-floating molluscs with chambered cells – came into their own, evolving more than 1000 genera.

By the late Cretaceous, some 65 million years ago, the plant and animal communities were as rich as those of today's world. Indeed, the dinosaurs seemed to be on their way towards intelligence, with upright species evolving forepaws that were

1. Some 65 million years ago, a sauropod dinosaur watches the comet that will exterminate her and all her living relatives.

2. Mercury's face looks as battered as that of the Moon. Earth would look the same if geology and weather did not steadily re-form its surface.

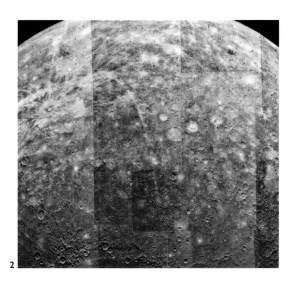

2

almost as mobile as hands and a stance that might, with further time, have become something oddly human. The first, shrew-like mammals evolved, but remained imprisoned in their few limited niches – in trees, down holes – until the tides of evolution turned to their advantage. Their chance came with the sudden and catastrophic death of the dinosaurs.

The impact theory

Throughout almost all the 20th century, scientists were puzzled by these two great hiatuses. What caused them? Few, except apparent eccentrics, considered the idea of catastrophic impacts. What is now considered hard evidence of an interplanetary bombardment was simply not recognized as evidence at all. The Moon's craters? They could be volcanic. Craters on other satellites

had not been seen in close-up. No one had seen any comet or asteroid colliding with anything. And where were the terrestrial craters, the smoking guns? True, there was one impressive scar – Meteor Crater in Arizona – but even in the mid-20th century there were those who argued that it was volcanic, on the grounds that no meteorites had been found.

In the late 1970s, answers began to emerge. The US lunar landing programme brought back rocks that had been compressed, showing that the Moon's craters were indeed made by impacts. The 1974–5 *Mariner* survey of Mercury, and then later surveys of outer Solar System satellites, showed cratering to be common.

And finally, scientists began to see that the Earth itself was not immune. Studies on Meteor Crater showed that the incoming meteorite, which struck about 50,000 years ago, would have vaporized on impact. Older, larger craters would have been largely eroded by weathering and ploughed over by the regular processes of continental drift and mountain building. More than 150 impact craters have now been identified. Thousands more must exist under rainforest canopies or polar ice, and many times that number would have been eroded away, not to mention the asteroids swallowed by the oceans. Almost all the craters are under 200 million years old, an indication that these represent a tiny fraction of actual impacts. Any figures for that period would have to be multiplied tenfold.

In 1980, Luis and Walter Alvarez capped theory with proof. This father and son team announced the discovery of an odd layer of clay near Gubbio, Italy, dating from the time the dinosaurs became extinct. It was rich in a rare element, iridium. According to models of terrestrial evolution, iridium should be concentrated in the Earth's core and in the primeval Earth's raw material, the asteroids – and in a particular type of asteroid, a chrondrite. The Alvarezes suggested that the only way the iridium could have become concentrated in this way was if the Earth had been hit by a huge meteorite 65 million years ago, and thus caused the extinction of the dinosaurs.

The Alvarezes' announcement caused a storm of controversy, but acceptance came gradually with the discovery of other 'iridium layers' in southern Spain, Denmark and New Zealand – over 50 sites in all by the mid-1980s – and further evidence of a catastrophic impact. In southern Spain, tiny, iridium-rich spheres of rock seemed to have been blasted from Earth into space, from where they fell back. In other layers of a similar date, fine soot particles were found, the evidence, apparently, of global forest fires. Then, in 1990, came the discovery of the smoking gun itself.

1. On an outcrop near Gubbio, Italy, a coin marks the rock-layer dividing the end of the Cretaceous period from the succeeding Tertiary period.

2. A reconstruction of the double ring of the 180-km (112-mile) Chicxulub Crater superimposed on today's coast of the Yucatán Peninsula.

▷ A SCALE FOR JUDGING THE RISKS

The 10-point Torino Scale is a chart for assessing the danger posed by an asteroid or comet. Created by Professor Richard Binzel of the Massachusetts Institute of Technology, it was adopted in 1999 by an international conference on Near Earth Objects held in Turin, Italy. The scale, which relates the size of the object and probability of collision, runs from 1 (small object, little chance of collision) to 10 (massive object, global catastrophe). So far, no asteroid or comet has been found that merits any rating at all.

Gravitational anomalies revealed the existence of a crater on the coast of the Yucatán Peninsula, deeply buried under younger sediments. In fact, the first evidence for the crater dates back to the 1940s, when the Mexican national oil company started to drill near the little town of Chicxulub (pronounced 'chic-sa-lube'). As often in the history of scientific advance, the evidence made sense only when enough other research had accumulated to provide a context within which to understand it.

Now, at last, scientists were able to explain what had happened. A 10-km (6-mile) asteroid had struck, blasting out a crater at least 180 km (112 miles) across, with catastrophic worldwide effects. An asteroid this size, travelling at 40,000 kph (25,000 mph), would have released 10,000 times as much energy as mankind's total nuclear arsenal. Acid rain poisoned the top 90 m (300 ft) of ocean. A tidal wave, 300 m (1000 ft) high, swept across the proto-Atlantic. The detritus blasted into orbit, falling as fireballs to ignite forests worldwide. A pall of dust blanketed the planet. It was a combination that slew all the large animals, and more than half of the smaller ones.

2

THE REVOLUTION COMPLETED

If there has been one impact, then surely there have been others: indeed, five major extinction events are now equated with asteroid impacts. But five major impacts over 1000 million years hardly points to a frequent bombardment, let alone any immediate danger. Do we really need to worry?

The answer, increasingly, is 'yes' because there is growing evidence of impacts in historical times. The evidence comes from three areas: tree rings, ice cores and historical records. This evidence suggests that folktales, once regarded simply as myths, should also be seen as indicators of real events.

From the 1920s, scientists have been building a picture of past climates from tree rings. This speciality, known as dendrochronology, is based on the fact that each year a tree's rate of growth depends on how good or bad the conditions are. Scientists now have year-by-year chronologies from widely scattered sites and several different species – Arizona pines, Californian bristlecone pines and European oaks. This evidence receives independent backup from Greenland's ice cap, which is made of snow compressed into ice layers over the past 40,000 years. The layers, which can be drilled out in long cores, record not only snowfall but also its chemical constituents.

Tree rings and ice cores both suggest that the world has gone through at least five climatic crises, which lasted several years. These dates centre on the years 1628 BC, 1159 BC, 207 BC, 44 BC and AD 540. The most likely cause, based on analysis of the effects of volcanic eruptions, is dust in the atmosphere from

Encounters and disasters

Date	Related events
1628 BC	Stonehenge abandoned. Exodus of Moses. Plagues of Egypt. End of Xia Dynasty in China. Eruption of Santorini.
1159 BC	Troy falls. End of Mycenaean civilization in Greece. Famine in Egypt. End of Shang Dynasty in China.
207 BC	Stones fall from sky in Europe. In China, famines mark end of Qin Dynasty and start of Han.
44 BC	Comet associated with death of Caesar. Famines in China.
AD 540	Dark Age starts in northern Europe. Plagues in Near East. Famines in China.

Flies plague Egypt in 1628 BC.

volcanoes, or dust from the tails of passing comets, or actual impacts. Or the cause may be all three.

With these dates in mind, histories and legends from cultures as far flung as China and Ireland, take on new significance. Around these dates the Sun was reported as pale, summers were non-existent, ashes fell from the sky, angels wielded 'heavenly swords', gods battled dragons, waters swept in from the ocean, lakes broke their banks, and plague and famine stalked the land. Many of these statements, formerly seen as mere legends, could be attempts to describe the appearance and effects of comets by people with no knowledge of the nature of comets, or of their cause-and-effect link with tectonic and atmospheric events – volcanoes, earthquakes, dust-clouds, or tidal waves (tsunamis).

To take one particularly dramatic example: tree rings and ice cores agree that 1628 BC was a particularly harsh year. This could have been the result of dust from a close comet, or from the eruption of Santorini, which annihilated the Mediterranean island in days, and could have provided Moses with his 'pillar of fire' during the Exodus from Egypt. The date also coincides with the start of extended disasters across the world. In China, a seven-year drought brought the end of the Xia Dynasty. Ireland suffered a seven-year drought too. At present the explanations for the evidence are highly controversial, but if they withstand analysis they will force a drastic revision of ancient world chronologies.

Similar events surround the year AD 540, when (among other things) Chinese 'dragons wrestled in ponds, and where they passed the trees were broken'; when earthquakes were recorded in Constantinople along with a great comet; and when English legends speak of a 'wasteland' following the death of King Arthur.

 DARK AGE OVER BRITAIN

Tradition relates that Britain suffered a grim time following the death of King Arthur (right) in the 6th century. A work said to be by St Patrick talks of Satan falling like a huge rock, and of walking for a month through a deserted and ravaged countryside. Tree rings back tradition, recording bitter conditions around AD 540. All these effects could have been caused by the aftereffects of a cometary impact or near-impact, and the release of dust that veiled the Sun and brought on a 'nuclear winter'. Another theory, suggested by the apparently intense local effects of the catastrophe, proposes an impact in the Irish Sea. Legend may back the idea of cometary impact – traditionally, Arthur is associated with dragons and swords, both ancient images for comets.

Impact on Jupiter

Although the conclusions from the evidence are debatable, they received support from the most dramatic event ever witnessed in our Solar System, something that showed the world that comets do indeed strike planets. In mid-March 1993, three astronomers at the Palomar Observatory in California were nearing the end of a 12-year survey of asteroids and comets. The three – David Levy, Carolyn Shoemaker and her husband Eugene – were depressed. The skies were cloudy, and their last box of film had been partially exposed accidentally. Levy realized that although the edges of the film were ruined, the central patches should be untouched. Between cloud patches they got to work. Two days later, on 25 March, Carolyn Shoemaker spotted a smear on the film near Jupiter. 'I don't know what this is,' she said. 'But it looks like a squashed comet.'

It was indeed. Intensive research into the comet's orbit showed that Shoemaker-Levy, as it was later named, was the remnants of a comet perhaps 10 km (6 miles) across captured by Jupiter and torn apart into a 'string of pearls' by its gravity in 1992. Moreover, the fragments had actually been captured by Jupiter and would strike the planet in July 1994.

Astronomers everywhere, briefed on the Internet, were agog. Witnessing an impact on Jupiter would be a hugely important experience, revealing new data about the gaseous giant and about the role of comets and asteroids in the evolution of the Solar System, and its life forms. Here was a once-in-a-millennium chance to observe an actual impact, with wonderful new

1. Eugene and Carolyn Shoemaker, co-discoverers of the comet named after them.

2. David Levy, third discoverer of Shoemaker-Levy, joined the mission to record threatening asteroids and comets.

tools – the Hubble Space Telescope and the *Galileo* spacecraft, then *en route* for Jupiter.

For over a year, the tension grew. Some scientists feared a non-event, a mere fizzle, while tabloid papers revelled in predictions of a cataclysm. Doubts and hype dissolved on 16 July, when the impact occurred. By then the comet had been reduced to 21 fragments ranging from 100 m (330 ft) to about 4 km (2.5 miles) across and stretching over 1.5 million km (1 million miles). The first one produced an awesome

3

3. The 'string of pearls' created when Jupiter tore apart Shoemaker-Levy 9.

4. Three of the 21 impact sites on Jupiter.

Following page: In an artist's reconstruction, a Shoemaker-Levy fragment strikes through Jupiter's upper atmosphere. The 'crater' below is a hole torn in the lower atmosphere by a preceding fragment.

4

explosion. It struck on the outer edge of the planet, just out of sight from the Earth, moving at 60 km (40 miles) per second. A fireball burst in Jupiter's shadowed atmosphere, and exploded upwards for 3000 km (1865 miles) into sunlight after only five minutes. Over the next 20 minutes, it ballooned outwards into a plume 10,000 km (6200 miles) across, producing an immense dark scar, the size of the Earth, which swung into full view as Jupiter spun. For a week, until 22 July, the rest of the fragments rained down on to Jupiter, some dissipating with no explosion, others blasting a line of immense dark spots into the atmosphere. These clouds, the dusty remains of the fragments, slowly grew fainter.

After eight months, scarcely a trace remained of the grandest event ever witnessed in the Solar System, providing evidence that will take many years to analyse. One lesson no one could doubt: the Earth evolved in a shooting gallery, and at some point will again be a target for bombardment.   ▷▷

A PRECARIOUS FUTURE

Today's 'celestial police force' is the Minor Planet Center (MPC) at the Smithsonian Astrophysical Observatory in Boston, USA, and they have given themselves a formidable task: to discover and track as many asteroids as possible, in particular Near Earth Objects (NEOs) or Earth Approachers. The MPC's job is to coordinate original observations from about 30 astronomers in four main US research centres and some 200 amateurs, combining these with follow-up results from two observatories in the Czech Republic.

   With special equipment to find moving spots of light and computers to process the results, progress has been astonishing. In records that are updated daily, the MPC tracks some 25,000 asteroids with known orbits and 30,000 whose orbits are less well established. A further 10,000 sightings await analysis. The number is climbing sharply, currently by 30,000 a year. The Center may receive reports of 1000 orbits a day, of which 50 per cent may be new. In a record month in 1999, some 5000 new objects were reported. Since Earth will eventually again be a target, the MPC and its related groups are determined to identify and track asteroids that may be a threat, in particular the 2000 NEOs of 1 km (0.6 miles) or more across.

The task ahead

So far the results are as nothing compared to the task ahead. The MPC's ambition is to log all the NEOs larger than 50 m (165 ft) that may be a threat

for the next century. Gareth Williams, the MPC's deputy director, estimates that there are 320,000 objects of 100 m (330 ft) across or more – and a million of 50 m (165 ft) across. It is a task that will take something like 20 to 25 years. Even then, a smaller threat will remain: a 10-m (33-ft) object would cause significant damage, and there are an estimated 150 million of these little threats.

   And NEOs form just a fraction of those in stable orbits. The MPC's wider aim is to record as many of these as possible. How many asteroids are there to be recorded? The MPC assumes that NEOs are a mere 0.001 per cent of the total: 3000 million asteroids are 50 m (165 ft) across or more. The original celestial police would have been astonished.

1

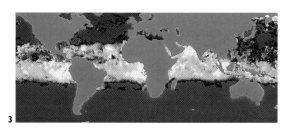

1. Eros approaches Earth. This 20-km (12-mile) wide asteroid is not a threat at present, but could be in the remote future.

2. and 3. How the 1991 volcanic pollution from Mount Pinatubo in the Philippines spread in days (top), and in two months (bottom).

Although future impacts are inevitable in the long run, the timing is totally unknown. No known rock is on a collision course with Earth, and all that astronomers can do for the next few years is gather information and sample the past to build a statistical analysis. According to current estimates, an impact as big as the one that struck Tunguska in Siberia might happen once a century, while bodies up to 1.6 km (1 mile) or so across might strike once every few hundred thousand years. Possibly, on the basis of the Permian and Cretaceous extinctions, planetesimals up to 10 km (6 miles) across might strike every 150 million years or so. But even a 'little'

strike by an asteroid, a mere 350 m (1150 ft) across, would release more energy than is stored by all the world's nuclear weapons.

The risks can be quantified in insurance terms. A report by the US Congress in 1992 estimated that an impact by a 1.6-km (1-mile) wide asteroid, causing worldwide devastation, might kill a quarter of the world's population. Such an asteroid might put nearly 1000 times more dust into the air than Mount Pinatubo did in 1991, an eruption that cooled the Earth measurably for a year. If one of these objects strikes once in 500,000 years, there is an annual risk to every individual of 1 in 2 million. Computed over an average lifetime, everyone on Earth has a 1 in 25,000 chance of dying by asteroid strike. For comparison, Americans have a 1 in 20,000 chance of dying in an aircraft crash, and a 1 in 60,000 chance of dying in a tornado.

How to clothe statistics with hard fact is the problem now facing astronomers and, increasingly,

✪ A comet strike could be catastrophic for mankind, but it would no more affect the progress of the Earth in orbit than the impact of a gnat affects a speeding truck.

policy-makers. Astronomers have a particular interest in the Taurid meteor shower, which arrives every 28 June (not a spectacular one because the shower occurs in daylight). The specks of dust lie in the same orbit as Encke's Comet, which circles the Sun relatively closely, within the confines of the Solar System. But there is more than dust in Encke's orbit: it is also associated with ten asteroids, and at least one astronomer – Duncan Steel of the Anglo-Australian Observatory – theorizes that the dust, the asteroids and the comet are all derived from a single large object that fell into the Solar System some 20,000 years ago. He places the risks of major impacts considerably higher than many of his colleagues – a Siberia-like strike every 50 years, a 1.6-km (1-mile) wide one every 100,000 years.

Officials must now deal not only with the facts as presented by astronomers but also with a need to respond to growing public unease. In the US, the House of Representatives first expressed concern in 1990, and directed NASA to intensify its detection programme. With no immediate threat to go on, officials have a problem. After all, if anyone in 1993 had asked: 'What are the chances of 21 objects smashing into Jupiter next year?' the answer would have been: 'One chance in many millions.' Yet within a year it happened. In Duncan Steel's words: 'Probabilistic arguments cannot be used in our present state of ignorance.'

What is to be done? As yet, no official plans exist because dangers emerge only to vanish again. With increasing frequency, potentially threatening

1. Kitt Peak, Arizona, is the headquarters of US Spacewatch, which paved the way for the wider-ranging, international Project Spaceguard.

2. US Spacewatch and the international Project Spaceguard co-operate to track the millions of objects like these that may threaten Earth in the future.

2

the appearance of something much larger. If that is all it is, it would be no threat at all. Anyway, further analysis of the orbit showed that it would come no closer than 4.2 million km (2.6 million miles) – 11 times the Earth–Moon distance. As a result, 2000 SG344 receded again into insignificance and was removed from the Torino Scale.

Still, the ammunition is out there. In 1994, James Scotti, part of the US Spacewatch programme set up at the University of Arizona's Steward Observatory, recorded an asteroid passing only 105,000 km (65,000 miles) away. In the long run, and possibly the short run, astronomical research and statistical analysis will have practical effects. Contingency plans will be made, crisis committees set up, money spent. Unless there is a major impact in the next decade, it seems likely that *Homo sapiens* is about to enter a new era of evolution. Earth now has a means of self-defence. As Hollywood has discovered with delight, the technology, in the form of rockets, guidance systems and explosives, already exists to blast or guide an asteroid from its destructive path. According to Tom Gehrels at the University of Arizona, the chances are that we would have a century to ward off any big strike, and 'given that time, a modest chemical explosion near an asteroid might be enough to deflect it'.

For the first time, a terrestrial species is in a position to secure its own survival in the face of a threat that has extinguished so many of its predecessors. The objects that created the world in which we evolved also have the power to destroy us; but they have also inspired the research that should, with luck, ensure our immortality.

asteroids are spotted, their orbits plotted and apocalyptic announcements made – only for research to show that there is no danger after all.

One recent example emerged in late 2000. Researchers in Hawaii recorded a 30–70-m (100–230-ft) object in a near-Earth orbit, which an Italian astronomer, Andrea Milani, suggested could impact the Earth in 2030. NASA put the chance of impact at 1 in 500, which gave the object, designated 2000 SG 344, the distinction of becoming the first to acquire a rating on the newly adopted Torino Impact Hazard Scale (▷p. 82). On 3 November 2000, the International Astronomical Union went public with the information, despite the fact that no one knew what the object was. Possibly, it is a piece of space junk, such as the fourth stage of an *Apollo* moon rocket, its high reflectivity giving it

FURTHER INFORMATION

BOOKS

Duncan Steel, *Rogue Asteroids and Doomsday Comets* (John Wiley, 1995). Steel is a renowned authority on comets. The foreword is by Arthur C. Clarke.

Walter Alvarez, *T. Rex and the Crater of Doom* (Princeton University Press, 1997). The story of the impact that killed off the dinosaurs, by one of the team who solved the mystery.

Mike Baillie, *Exodus to Arthur: Catastrophic Encounters with Comets* (Batsford, 1999). The evidence for past cometary impacts, revealed by tree-rings and ice-cores.

Gerrit Verschuur, *Impact! The Threat of Comets and Asteroids* (Oxford University Press, 1996). A survey by a noted American radio astronomer.

David Levy, *Comets: Creators and Destroyers* (Simon & Schuster, 1998). An account by one of the discoverers of the Shoemaker-Levy comet that hit Jupiter in 1994.

Sara Schechner, *Comets, Popular Culture and the Birth of Modern Cosmology* (Princeton University Press, 1997). A detailed account by a leading historian of astronomy.

Jacques Crovisier and Thérèse Encrenaz, *Comet Science: The Study of Remnants from the Birth of the Solar System* (Cambridge University Press, 2000). A summary of cometary research, with much to interest the non-specialist.

J. Kelly Beatty, Carolyn Collins Petersen and Andrew Chaikin (eds) *The New Solar System* (Cambridge University Press). A detailed yet readable survey of the subject, with 28 chapters by acknowledged experts. Includes chapters on the Kuiper Belt, the Oort Cloud, comets, asteroids, meteorites and the role of impacts (by Eugene and Carolyn Shoemaker).

Donald Yeomans, *Comets: A Chronological History of Observation, Science, Myth and Folklore* (Wiley, 1991). The most detailed of surveys for non-specialists.

Carl Sagan and Ann Druyan. *Comet* (Random House, 1985). Still one of the finest accounts, with numerous illustrations.

WEBSITES
NASA
http://www.nasa.gov

Spacewatch
http://www.lpl.arizona.edu/spacewatch

The Spaceguard Foundation homepage
http://spaceguard.ias.rm.cnr.it.SGF

Spaceguard UK and the Sainsbury Task Force report
http://www.nearearthobjects.co.uk

The Minor Planet Center and near-earth objects
http://cfa-www.harvard.edu/cfa/ps/mpc.html

The International Astronomical Union homepage:
http://www.iau.org

INDEX